REWARD

REWARD

Pre-intermediate

Practice Book

**Liz Driscoll
Simon Greenall**

MACMILLAN
HEINEMANN
English Language Teaching

READING

1 Do foreign visitors ever use English when they come to your country? When?

2 Read *What's in a name?* What do you think is the most important piece of advice in the text?

What's in a name?
....

The first exchange between two people – and the first chance to make a good or bad impression – is often an exchange of names.

In the Eastern Hemisphere, name often shows social or family status and a mistake can be an insult. Using someone's first name before the person gives permission can also be very rude.

'What shall I call you?' is always the first question of one director of an international telecommunications corporation. 'It is better to ask many times,' he advises, 'than to get it wrong.' Even then, he says 'I treat people formally until they say "Call me Joe".' Another world traveller always studies a list of important people he will meet, country by country, surnames underlined, on the flight there.

The next question is: How do you know which name is the surname? In China and Thailand the surname comes first and the first name comes last. But the Thais use *Mr* with the first name and not the surname. The Taiwanese often have an extra first name before any of their other names – the polite way to address someone with the full name Tommy Ho Chin is Mr Ho. The Japanese are usually very formal, and you address them in the same way that they address you. When a Japanese person uses your first name, add *san* to his or her name. Don't use *Mr* or *Mrs*.

A bit complicated? The best thing to do is to ask.

3 Write the full names of four of your friends or family. Circle their first names and underline their surnames.

4 Are there any special rules for using surnames and first names in your country?

SOUNDS

Say these sentences aloud and underline the stressed words.

1 Pleased to meet you.

2 How old are you?

3 How do you do.

4 What's your first name?

5 Can I help you?

Now listen and check.

VOCABULARY

1 Complete the crossword.

Across:
1 Can I _____ you? (4)
4 _____ do you live? (5)
5 I don't speak English at home with my _____ . (6)
7 How _____ you? (3)
8 Pleased to _____ you. (4)
10 What's your _____ name? (5)
11 How do you _____ . (2)
13 I'm sorry. I don't _____ . (10)

Down:
1 _____ , Bill. How are you? (5)
2 I'd like a Coke, _____ . (6)
3 Could you _____ that? (6)
5 Are you _____ the USA? (4)
6 I haven't got a husband. I'm not _____ . (7)
9 How much _____ this? (2)
12 How _____ are you? (3)

Crossword answers filled in:
- 1 Across: HELP
- 3 Down: R E P E A T
- 4 Across: WHERE
- 5 Across: FAMILY
- 5 Down: FROM
- 7 Across: ARE
- 8 Across: MEET
- 9 Down: IS
- 10 Across: FIRST
- 13 Across: UNDERSTAND
- 2 Down: PLEASE
- 6 Down: MARRIED
- 11/12 Down: DO / OL

2 Match the pairs of sentences. Put the correct letters in the boxes.

1 [e] 2 [c] 3 [a] 4 [b] 5 [d]

1 How do you do.
2 How are you?
3 This is Anna.
4 Come in.
5 Can I help you?

a Pleased to meet you.
b Thank you.
c Fine, thanks. And you?
d Yes, please. How much is this?
e How do you do.

3 Find fifteen verbs in the puzzle. They go in two directions: → and ↓. Use each letter once only.

W	A	N	T	S	I	N	G
A	T	S	T	O	D	A	A
R	H	T	A	F	R	S	C
R	I	A	K	F	I	K	C
I	N	Y	E	E	N	S	E
V	K	G	O	R	K	I	P
E	V	I	S	I	T	T	T
L	I	V	E	T	A	L	K

GRAMMAR

1 Complete these sentences with six of the verbs in the puzzle in *Vocabulary* activity 3.

1 We often *visit* friends on Saturdays.
2 We usually *take* a small gift for our host or hostess.
3 We don't usually *talk* about religion or politics.
4 We don't usually *ask* personal questions.
5 We usually *drink* coffee at the end of a meal.
6 We don't often *go* to restaurants with friends.

2 Choose four of the other verbs in the puzzle in *Vocabulary* activity 3. Write sentences saying what you do and don't do in your country.

1 _____
2 _____
3 _____
4 _____

3 Put the words in order and make sentences about what this person does or doesn't do when she accepts hospitality.

1 the never dishes I wash to offer
I never offer to wash the dishes.
2 ten I minutes always late about arrive
I always arrive about ten minutes late
3 I take or wine usually chocolates
I usually take chocolates or wine
4 coffee I after sometimes meal want the
I sometimes want coffee after the meal
5 to my dinner I often own parties don't on go
I often don't go on my own to dinner parties
6 smart I wear clothes don't usually
I usually don't wear smart clothes

4 Complete this conversation.

YOU *Are you American?*
PETER Yes, I come from New York City.
YOU *Where do you live?*
PETER I live in London.
YOU *What do you do*
PETER I work in computers.
YOU *Are you married*
PETER Yes. My wife's English.
YOU *What's your name*
PETER Delgado. Peter Delgado.

2 | *A day in the life of the USA*

VOCABULARY

1 Write the times shown on the clocks in two ways.

1

seven pm
seven o'clock in
the evening

2

one fifteen am
a quarter past one
in the morning

3

two forty five pm
a quarter to three
in the evening

4

nine thirtee am
half past nine in the
morning

5

eleven forty five pm
a quarter to twelve in
the evening

6

eight ten am
ten past eight in the
morning

2 Put these things in the order you do them by numbering the boxes. Then complete the sentences with the times.

- [6] I come home *at half past four pm*
- [1] I wake up *at seven am.*
- [9] I have dinner *at eight p.m*
- [7] I get ready for school/work *at seven pm*
- [2] I get up *at ten past seven am*
- [10] I go to sleep *at half past nine pm*
- [5] I have lunch *at half past twelve am*
- [4] I leave home *at eight pm*
- [3] I have breakfast *at twenty past seven am*
- [8] I watch TV *at seven pm*

SOUNDS

1 Say these words aloud. Underline the three words with the same-sounding ending.

1 comes goes has makes
2 gets sings stops works
3 dances finishes leaves washes
4 asks does sits takes
5 arrives dresses refuses watches
6 lives offers serves wakes

Now listen and check.

2 Listen and say the words again.

GRAMMAR

1 Complete these sentences with one of the words in brackets.

1 My sister *goes* shopping after work. (go/goes)
2 My brother *works* in a bar. (work/works)
3 What time does your teacher *finish* work? (finish/finishes)
4 The teacher never *sits* in class. (sit/sits)
5 My sister *doesn't* live at home. (don't/doesn't)
6 George and Elsie *watch* TV after dinner. (watch/watches)
7 Do your children *offer* to help you much? (offer/offers)
8 We *don't* have lessons at night. (don't/doesn't)

4

2 Write true sentences about yourself or people you know. Use the other eight words in brackets in activity 1.

1 _____
_____

2 _____
_____

3 _____
_____

4 _____
_____

5 _____
_____

6 _____
_____

7 _____
_____

8 _____
_____

3 Complete this description of Annie Laurence's day with suitable words. Then check your answers in your Student's Book.

7.15am Roanoke, Virginia. A tired Annie Laurence, 10, wakes (1) _up_ and gets ready for school. An hour later she leaves (2) _home_ . She has (3) _lunch_ at school, sandwiches (4) _and_ an apple. It's a long day for Annie. She (5) _doesn't_ get home again until 5pm at the end of the (6) _day_ .

4 Complete this description of the rest of Annie Laurence's day with suitable verbs.

Annie Laurence (1) _gets_ home at 5pm after a long day at school. She usually (2) _has_ a soft drink and plays with her sister. Dinner is about an hour later. After dinner, Annie (3) _washes_ up with her mother and then she (4) _watches_ TV for an hour. At eight o'clock, she (5) _gets_ ready for bed. She (6) _goes_ to sleep at about half past eight.

LISTENING AND WRITING

1 🔲 Listen to one of the people in the photos talking about her day. Who is speaking?

2 What does the speaker do at these times?

8.30am _____
12.30pm _____
6.00pm _____
10.30pm _____

3 What else does the speaker do during the day? Make notes.

🔲 Listen again and check.

4 Write a description of the speaker's day.

5 Choose someone you know and write a description of his or her day.

3 | *Home rules*

GRAMMAR

1 Complete the paragraph with *a/an*, *the* or put –
if there's no article.

Larry and Belinda Hope live in (1)_—_ Keswick,
(2)_a_ small town in (3)_the_ north-west of (4)_—_
England. Larry is (5)_a_ writer and he works at
(6)_—_ home. One of (7)_the_ bedrooms is now
(8)_an_ office.

LARRY 'It's (9)_—_ my favourite room in (10)_the_
house – I love it! It's got (11)_an_ old desk and
(12)_a_ couple of big, wooden cupboards. I've
got (13)_a_ lots of books and there's (14)_an_
enormous bookcase next to (15)_the_ desk. Then
there's (16)_a_ sofa-bed – that's really for
visitors but sometimes I sleep on it in (17)_the_
afternoon. This room has also got (18)_a_ door
into (19)_the_ back garden and when (20)_the_
weather's good, I often work outside.'

2 These are your answers. Write A's questions
about the rooms in your home.

1 **A** Is there a kitchen in your house?

YOU Yes, there is.

2 **A** Is there a sofa-bed in your room?

YOU No, there isn't.

3 **A** Are there any animals in your bedroom?

YOU Yes, there are.

4 **A** Are the any visitors/lamps

YOU No, there aren't.

3 Complete these sentences and write about your
home.

1 There's a bed and a lamp in the bedrooms

2 There isn't a lamp in the bathroom

3 There are chairs in the kitchen —

4 There aren't books in the toilet, there
aren't animals in the bathroom

4 Write the plural forms of these words.

singular	plural
1 bus	buses
2 curtain	curtains
3 class	classes
4 family	families
5 party	parties
6 play	plays
7 sandwich	sandwiches
8 shoe	shoes
9 window	windows
10 woman	women

SOUNDS

Say these sentences aloud and underline the
stressed words.

1 Do you live in a house or a flat?
2 What's the main room in your home?
3 How many bedrooms are there?
4 Have you got a dishwasher?
5 Where do you eat your meals?
6 Do you have a television?

📼 Listen and check.

LISTENING AND WRITING

1 📼 You are going to hear a woman answering the questions in *Sounds*. Listen and tick (✓) her answers.

1 house ❑ flat ❑
2 living room ❑ kitchen ❑
   bathroom ❑ bedroom ❑
3 one ❑ two ❑
   three ❑ four ❑
4 yes ❑ no ❑
5 living room ❑ kitchen ❑
6 yes ❑ no ❑

2 What else does the woman say about her home? Make notes.

_____
_____
_____
_____
_____
_____
_____
_____
_____
_____
_____

📼 Now listen again and check your answers to activities 1 and 2.

3 Answer the questions in *Sounds* about your own home.

1 _____
   _____
2 _____
   _____
3 _____
   _____
4 _____
   _____
5 _____
   _____
6 _____
   _____

READING AND VOCABULARY

1 Read these short extracts which each include the word *home*. Decide what the expressions with *home* mean in these contexts. How would you say these expressions in your own language?

1 'I was born in Northern Ireland but I came over to England to do my training – I'm a nurse. England's all right, I suppose, but I like to go **home** every now and then.'

2 'My gran lived on her own for a bit after grandad died. But she's getting on a bit now – she's 84, actually – so she moved into a **home** last year. She's looked after there so my parents don't have to worry about her so much.'

3 'I'm quite happy to stay **at home** most evenings. I'm out at work all day so I'm in no hurry to go out in the evenings as well.'

4 'My brother's football-mad. He's a Manchester United supporter and he's got all the players' autographs. He goes to all their matches when they're playing **at home** and he even goes to some of the away ones.'

5 'My flat seems to get smaller as the years go on. I've got so many things now and I never seem to be able to **find a home** for them.'

6 'A male friend of mine is an absolutely wonderful cook. He makes the most fantastic pasta dishes you can imagine. He's really **at home** in the kitchen.'

7 'When I worked in Germany as an au pair, I had a great time. The family I worked for were very nice and they treated me like one of them. It was really a **home from home**.'

2 Which of these expressions are most useful to you?

READING AND WRITING

1 Look at the words below. Which things do you like or dislike?

television   radio   winter   instant coffee   queues
sausages   parks   football   newspapers   gardening

2 Read these quotations by British people talking about something they like or dislike. What are they describing? Choose from the list in 1.

1
> 'There's so much variety: plays, quiz shows, news, interviews, documentaries and of course the cricket commentary. You really need your imagination, it's much better than television.'
> *Pamela, London*

2
> 'I am always surprised to find how much green space there is in the average English city. I really appreciate being allowed to walk on the grass, especially as I don't have a garden of my own. And there's something for everyone to do there.'
> *Frances, Leicester*

3
> 'I am always astonished how twenty-two grown men can rush around chasing a ball, shouting at each other and generally behaving like children. Surely they've got better things to do.' *Edna, Durham*

4
> 'Well, it's our sense of fair play, I think. It means that the person who arrives first gets served first or gets on the bus before everyone else. That seems fair to me. I think it's a sign of a civilised country.' *Henry, Bradford*

5
> 'It goes on for ages, from the end of October when it gets dark early, until March or sometimes April. Round about February I have to get away and find some sunshine. I get so depressed by it.' *Sally, Brighton*

6
> 'I get one every day but Sunday is special. For me, Sunday morning means getting up late, going down to the newsagents and buying two or three, then making breakfast and spending the rest of the morning drinking coffee and reading them. Lovely!' *Bridget, Henley*

7
> 'Well, most people who live in a house have somewhere at the back which they plant with flowers or vegetables, but I really can't be bothered. My back yard is an absolute mess, and I prefer it that way.' *Ken, Farringdon*

3 Write sentences saying what the people like or dislike and if you agree with them.

4 Think of something you like about your country and write a short paragraph explaining why. Try not to mention what it is.

SOUNDS

1 🔲 Listen and underline the words with two syllables.

<u>awful</u>   beer   cheap   cold   crowded   dirty
driving   food   friendly   great   hotels   litter
parks   police   polite   shopping   slow   tourists
warm   weather

Now say the words aloud.

2 🔲 Listen to these sentences. Put a tick (✓) if the people use a strong intonation for something they like or dislike a lot.

1   I can't stand shopping.
2   I hate British winters.
3   I love walking in the country.
4   I don't like rock music at all.
5   I like playing tennis very much.
6   I hate going to parties.

Now say the sentences aloud. Use a strong intonation.

VOCABULARY

Complete these sentences with words from *Sounds* activity 1.

1   I don't like the underground. It's very
    crowded during the rush hour and you
    don't usually get a seat.
2   I don't usually hire a car when I'm in Britain
    because I don't like _driving_ on the left.
3   I love eating foreign _food_ – Italian
    and Chinese are my favourites.
4   Irish people are really _friendly_ – they
    always stop to talk.
5   I don't like seeing _litter_ on the streets.
    Why don't people take their rubbish home?
6   A good way to get round Turkey is on public
    transport – buses and minibuses are both
    frequent and _cheap_ .

GRAMMAR

1 These are your answers. Write nine different questions for your answers.

1 Do you watch T.V.?
Yes, I do. A little.

2 Do you wear jeans?
Yes, I do.

3 Do you like chocolate?
Yes, I do. A lot.

4 I lost my ball?
It's all right.

5 Do you like jazz?
I don't mind it.

6 Do you go swimming?
No, I don't. Not very much.

7 Do you like the noise?
No, I don't.

8 Do you learn the example?
Not at all.

9 Do you like the pollution?
I hate it.

2 Add -ing words and phrases to Paul's sentences about his likes and dislikes.

1 I like my friends.
I like visiting my friends.

2 I love discos.
I love going to discos

3 I like breakfast in bed.
I like having breakfast in bed

4 I don't mind jazz.
I don't mind listening to jazz

5 I like parties.
I like having parties

6 I don't mind shopping.
I don't mind going shopping

7 I love champagne.
I love drinking champagne

8 I hate tennis.
I hate playing tennis

3 How do you feel about the underlined things in activity 2? Use the underlined words and write eight true sentences.

1 _____
2 _____
3 _____
4 _____
5 _____
6 _____
7 _____
8 _____

4 Write your responses to what A says.

1 A I like classical music.
YOU _____

2 A I don't mind washing up.
YOU _____

3 A I hate getting up early.
YOU _____

4 A I love Italian food.
YOU _____

5 A I don't like cold weather.
YOU _____

5 Write your responses to what this student says.

1 'I don't like listening to tapes.'
_____

2 'I don't mind working in pairs.'
_____

3 'I love writing in English.'
_____

4 'I like doing grammar exercises.'
_____

SOUNDS

1 🔊 Listen to the sentences and underline the stressed words.

1 She's waiting for someone.
2 He's doing the shopping.
3 They're standing by the road.
4 He's playing an accordion.

2 Say these sentences aloud and underline the stressed words.

1 He's painting a picture.
2 They're walking to work.
3 She's thinking about her friend.
4 He's holding something.

🔊 Now listen and check.

GRAMMAR

1 Write the infinitive forms of these *-ing* form verbs.

-ing endings	infinitive
1 asking	ask
2 carrying	carry
3 going	go
4 watching	watch
5 coming	come
6 leaving	leave
7 shining	shine
8 writing	write
9 getting	get
10 sitting	sit
11 stopping	stop
12 travelling	travel

2 Complete these sentences with the present simple and present continuous form of the same verb.

1 My brother usually _wears_ jeans and a sweater but he _is wearing_ a suit today.
2 Tony _works_ for the Disney Corporation in Long Beach. He _is working_ with another imagineer this afternoon.
3 I _drink_ a lot of coffee but I _am drinking_ a glass of wine right now.
4 I _am having_ lessons with a local driving school. I usually _have_ three lessons a week.
5 Shirlee never _watches_ TV during the day. She _is watching_ the news at the moment because her children are asleep.
6 I _read_ every night in bed and I _am reading_ a really good book at the moment.

3 Write six sentences about yourself. Use six verbs from activity 2 in the present simple or the present continuous.

1 I wear cheap clothes.

2 I am writing a letter for my uncle and my cousin.

3 I am drinking a cup of tee

4 I usually watch T.V. on saturday.

5 I read every night in bed.

6 For every picnic, I sit on the grass

VOCABULARY

1 Are the underlined words in these sentences nouns or verbs? Write N or V in the boxes.

1 The <u>play</u> starts at 7.30 pm. [N]

2 I <u>work</u> in a bank. [ ]

3 People <u>queue</u> for tickets at weekends. [ ]

4 (✓) is a tick and (✗) is a <u>cross</u>. [ ]

5 There's a <u>talk</u> about Fellini at the cinema club. [ ]

6 Let's go for a <u>walk</u> this afternoon. [ ]

2 Write sentences of your own with the six underlined words in activity 1. Use the verbs as nouns and the nouns as verbs.

1 *I don't play tennis every week.*

2 _____

_____

_____

3 _____

_____

_____

4 _____

_____

_____

5 _____

_____

_____

6 _____

_____

_____

READING AND WRITING

1 Choose sentences which describe what people are doing in four of the pictures. Write the number of the sentence next to the correct picture.

1 There's a man standing by the window, feeding the birds.

2 A young boy is playing a guitar.

3 There's a woman cooking dinner.

4 There's a man reading a newspaper.

5 A woman is watching television.

6 There are two people having something to eat.

7 There's a man having a bath.

8 An old woman is doing the washing up.

9 A couple are getting ready to go out.

10 A young woman is having a cup of tea.

11 A young man is just coming in through the front door.

12 There's a man going to bed.

2 Write sentences to say what people are doing in the other four pictures.

_____

_____

_____

_____

SOUNDS

Say these words aloud. Underline the three words in each group with the same-sounding ending.

1 expected   played   started   visited
2 called   finished   liked   walked
3 enjoyed   lived   looked   tried
4 danced   decided   stopped   watched
5 asked   continued   stayed   travelled
6 listened   opened   replied   wanted

[cassette icon] Now listen and check.

GRAMMAR

1 Use one word from each group in *Sounds* to complete these sentences.

1 I _started_ my homework at six o'clock.
2 Paul Theroux _walked_ around Britain.
3 They _lived_ in a flat near the station.
4 We _watched_ a film on TV.
5 The teacher _asked_ lots of questions.
6 My uncle _opened_ the door.

2 Use the negative form of one of the other words from each group in *Sounds* to complete these sentences.

1 We _didn't play_ tennis.
2 I _didn't like_ the man's scarf.
3 My friends _didn't look_ at my photos.
4 He _didn't dance_ with his girlfriend at the party.
5 They _didn't stay_ in a hotel.
6 She _didn't want_ a drink.

3 Find twenty verbs in the puzzle. They go in two directions: ↑ and →. Circle the infinitives and underline the past tense forms. Use each letter once only.

B	L	E	F	T	S	A	I	D
E	G	A	V	E	T	O	L	D
C	H	M	S	W	R	O	T	E
O	A	E	L	H	D	I	D	B
M	V	E	E	E	R	S	T	E
E	E	T	E	A	U	I	A	M
G	O	T	P	R	N	T	K	A
T	H	O	U	G	H	T	E	K
C	A	M	E	K	N	E	W	E

4 Write the past simple forms of the infinitive forms from the puzzle, and write the infinitive forms of the past simple forms from the puzzle.

past simple	infinitive
became	leave
had	says
met	give
spent	tell
heard	write
ran	do
sat	get
took	think
was/were	come
made	know

5 Are the underlined verbs in these sentences in the present simple or the past simple? Write PR (present) or PA (past) in the boxes.

1 [PA] That plane ticket <u>cost</u> a lot.
2 [PR] My father always <u>shuts</u> the front door.
3 [ ] I <u>read</u> about the plane crash in the paper.
4 [ ] I <u>put</u> the light off before I get into bed.
5 [ ] I <u>cut</u> my hand on some glass the day before yesterday.
6 [ ] These days most baseball players <u>hit</u> the ball quite hard.

6 Write sentences of your own with the underlined verbs in activity 6. Use the present simple verbs as past simple verbs and the past simple verbs as present simple verbs.

1 _____
2 _____
3 _____
4 _____
5 _____
6 _____

VOCABULARY

1 Use two words from the corresponding line to complete each sentence.

1 long   people   scarf   smart
2 blue   coat   flower   warm
3 flat   grey   shoes   sky
4 bag   big   heavy   smile
5 boat   blue   man   old
6 bunch   cloud   lovely   low

1 *Smart people* often wear gloves.
2 There was a _____ _____ in her hat.
3 Her _____ were good for walking and, in addition, the path was _____ .
4 The _____ in her right hand was small and not very _____ .
5 I saw an _____ _____ but he didn't see me.
6 A _____ black _____ passed in front of the sun.

2 Write sentences with the other two words in each line in activity 1.

1 _____
2 _____
3 _____
4 _____
5 _____
6 _____

LISTENING

1 In *The Kingdom by the Sea*, Paul Theroux describes a journey he made around Britain. Look at the map and find Deal, where he started his journey.

🔊 Here are some of the things Paul Theroux did on his journey. Listen to Emma, who has read the book, talking about the journey, and number the things Paul Theroux did in the order he did them.

☐ A man called Mr Bratby painted his portrait.
☐ He visited the famous British travel writer, Jan Morris.
☐ He stayed in a bed and breakfast for his first night.
☐ He nearly saw the Queen.
☐ He took a boat to Northern Ireland.
☐ He started his journey on May 1st.

2 Where did Paul Theroux do the things in activity 1? Write numbers in the boxes.

Brighton ☐   Criccieth ☐   Deal ☐
Hastings ☐   Liverpool ☐   London ☐
Margate ☐   Southend ☐   St Andrews ☐
Stranraer ☐

🔊 Now listen again and check.

13

The world's first package tours

SOUNDS

🔊 Listen to these questions. Put a tick (✓) if you think the speaker sounds interested.

1 How long was your holiday?
2 Did you have a good time?
3 What clothes did you wear?
4 Was it expensive?
5 Who booked the tickets?
6 Did you take any photos?

Now say the sentences aloud.
Try to sound interested.

VOCABULARY

1 Put the words in order and write sentences.

1 I airport taxi to took the a
   I took a taxi to the airport.
   _____

2 I the cheques my traveller's
   changed at bank
   _____

3 she cash handbag in her carried
   _____

4 we holiday year choose a package
   didn't last
   _____

5 receptionist a young checked
   coupons the
   _____

6 holiday airport us at our guide
   met the
   _____

2 Underline the objects of the sentences in activity 1.

LISTENING AND WRITING

1 🔊 Listen to four people talking about their holidays. They are each answering one of the questions in the list. Put the number of the speaker in the box by the question they answer.

☐ When was your holiday?
☐ Where did you go?
☐ Why did you choose that place?
☐ What was the weather like?
☐ Who did you go with?
☐ Which part of the holiday did you enjoy most?

2 Answer the questions for each speaker.

Speaker 1
_____

Speaker 2
_____

Speaker 3
_____

Speaker 4
_____

🔊 Now listen again and check.

3 Read some more about the four speakers' holidays. Look at the questions in *Sounds* and decide which ones they are answering. Put the numbers of the questions in the correct boxes.

☐ 'Two weeks, I think. We left on a Friday and came back on a Sunday. That's sixteen days.'

☐ 'We booked our own. I went from Heathrow and the other two went from Manchester because they live in the north.'

☐ 'No, not at all. It was really cheap. I didn't spend very much money at all.'

☐ 'Yes, I took hundreds and I've got copies of my friend's too. Do you want to have a look at them?'

🔊 Now listen and check.

GRAMMAR

1 These are your answers. Write questions about your last holiday.

1 _____
Yes, I did.

2 _____
No, I didn't.

3 _____
Yes, it was.

4 _____
No, it wasn't.

5 _____
Yes, it did.

6 _____
No, it didn't.

7 _____
Yes, there were.

8 _____
No, they weren't.

2 Do the quiz *Around the World*.

3 Use the information in the corresponding questions and answers in the quiz to complete eight different questions.

1 When *did the Titanic hit an iceberg?*
2 Who _____
3 What _____
4 Where _____
5 Who _____
6 When _____
7 What _____
8 What _____

4 Write four more questions for the quiz. Ask about the subject of the sentence.

1 _____
2 _____
3 _____
4 _____

AROUND THE WORLD QUIZ

Win the holiday of a lifetime! Win two air tickets for an AROUND THE WORLD trip. All you have to do is to answer these eight questions.
(The answers are at the bottom of the page.)

1 Which boat hit an iceberg and sank on its first voyage in 1916?
2 Where did Neil Armstrong go in 1969?
3 Who made the first solo flight across the Atlantic in 1927?
4 Which Italian went to China in the 12th century?
5 Where did Roald Amundsen go in 1912?
6 What happened in Nepal on 2nd June 1953?
7 When did Disneyland in California first open to the public?
8 Who wrote *Around the World in Eighty Days*?

1 ....
2 ....
3 ....
4 ....
5 ....
6 ....
7 ....
8 ....

5 Write four more questions for the quiz. Ask about the object of the sentence.

1 _____
2 _____
3 _____
4 _____

Quiz answers
1 the *Titanic*
2 the moon
3 Charles Lindbergh
4 Marco Polo
5 the South Pole
6 Sir Edmund Hillary and Sherpa Tenzing reached the top of Everest
7 1955
8 Jules Verne

15

8 | *Something went wrong*

SOUNDS

Match the words with the same vowel sound.
Put the correct numbers in the boxes.

1 saw	☐	got
2 cost	☐	gave
3 said	☐	hit → *happen*
4 came	1	thought
5 did	☐	left

🔊 Now listen and check. Say the words aloud.

GRAMMAR

1 Complete this description of what Mark did yesterday with suitable verbs. Use some of the verbs more than once.

I (1) *got* up early and (2) *had* ✓ a shower. Then I (3) *got* dressed and (4) *went* downstairs. I (5) *had* my breakfast in front of the TV.

After breakfast I (6) *read* some letters and (7) *went* to the post office on my bicycle. I (8) *had* lunch with my sister and her family. In the afternoon I (9) *took* my nephews to the cinema. We (10) *watched* a Walt Disney cartoon. It (11) *was* quite good!

In the evening two friends of mine (12) *came* here for a meal. I (13) *made* a chicken curry because I (14) *had* some chicken in the fridge. My friends (15) *had* a long journey home so they (16) *left* at about half past ten. I (17) *was* really tired.

2 Write sentences saying what Mark didn't do yesterday.

1 *He didn't stay in bed late.*
2 *He didn't have his breakfast in the kitchen.*
3 *He didn't go to the post office by car.*
4 *He didn't accompany her nephews for shopping.*
5 *He didn't take potatoes in the fridge.*
6 *He wasn't even excited.*

3 These are your answers. Write A's questions.

1 A _____
_____
YOU Last week.

2 A _____
_____
YOU Yesterday morning.

3 A _____
_____
YOU The day before yesterday.

4 A _____
_____
YOU Four days ago.

5 A _____
_____
YOU In 1992.

6 A _____
_____
YOU The month before last.

4 Find the endings for the sentences. Write the letters in the boxes.

1 I went home by taxi ☐☐
2 I slept badly ☐☐
3 I didn't enjoy the film ☐☐

a ...so I felt awful the next day.
b ...because I had a lot of things to carry.
c ...because I didn't think the acting was very good.
d ...because I ate my dinner quickly.
e ...so I got back in time for the phone call.
f ...so I switched it off and went to bed.

5 Complete the sentences in activity 4 in another way.

1 I went home by taxi
   a because _____
   b so _____

2 I slept badly
   a because _____
   b so _____

3 I didn't enjoy the film
   a because _____
   b so _____

6 Write a description of what you did yesterday. Use regular and irregular past tense forms. Join your sentences with *so* and *because*.

READING AND WRITING

1 Read these extracts from an Italian student's learning diary. Match the extracts with the activities in Lesson 8 of your Student's Book.

2 Write about some of the activities you did in Lesson 8 of this Practice Book. Try to use English but use your own language if you like.

_____
_____
_____
_____
_____
_____
_____

Lesson 8; Student's Book

☐ Then we looked at expressions of past time. I made some mistakes with the word 'last'. I put it after the noun, but in English it goes before the noun.

☐ Finally we wrote some class stories. We wrote a sentence and then gave the sentence to another student. This student read that sentence and then added the next sentence to the story. I enjoyed writing the stories because we had to read other people's sentences and then write something. The teacher said the activity was communicative. I think there were a lot of mistakes but the teacher didn't mind.

☐ After that we did some grammar exercises with the past tense. I looked at the list of irregular verbs at the back of the Student's Book.

☐ We started with a vocabulary activity. It wasn't very difficult because I already knew some of the words about travel. We worked in pairs. My partner and I talked in Italian but that was OK.

☐ Next we listened to two people telling stories. I'm not very good at listening to English but we listened to the tape many times so it was all right.

17

READING

1 Read *Family life*. What does Kathy actually tell you about herself? *elle même*

Kathy Papas, who lives in Sydney, Australia, talks about her family life.

'My parents came here from the Greek island of Kalymnos in 1965. I was eight years old, the eldest of four children. We shared a house with three other families from Kalymnos when we arrived in Sydney. We were quite lucky – both my parents found jobs and after about 20 months here they bought a house. But life wasn't easy for my parents. Language was a real problem and they never really settled down here. They missed their friends back home, Greek hospitality in general, and the closeness to nature they had on the island. In 1980 they moved back to Greece and my father died five years later.

My husband and I socialise almost entirely with our families – his sister and my brothers and their children – and other Greek-Australians. We meet in each other's homes and sometimes we have a barbecue together. Most Greeks live in the suburbs to the east of Sydney so we don't have far to travel to see each other.

We also send our three daughters to Greek classes twice a week after regular classes. They were all born here but a lot of Australians still tend to think of them as foreigners. This is perhaps the reason that we Greek-Australians stay together like a big family.'

2 Does she give the answers to these questions in the text? Write Y or N in the boxes.

1 Who's the head of the family? ☒N

2 How many people live in Kathy's home? ☒ N

3 How many brothers has she got? ☒ N

4 How often does she see her brothers? ☒✓

5 Is Kathy's family Greek or Australian? ☒N

3 What do you think are the answers to the questions in activity 2?

VOCABULARY

1 Write these words in the correct column of the chart.

1 aunt   2 boyfriend   3 daughter   4 father
5 girl   6 grandmother   7 husband
8 nephew   9 sister   10 woman

1	*aunt*	*uncle*
2	girlfriend	*boyfriend*
3	daughter	son
4	mother	father
5	girl	boy
6	grandmother	grandfather
7	wife	husband
8	niece	nephew
9	sister	brother
10	woman	man

2 Complete the chart in activity 1.

SOUNDS

🔊 Listen and underline the /ə/ sounds.

1 I don't know many children.

2 I haven't got any sisters.

3 I live with my parents.

4 I come from a small family.

5 I haven't got any cousins.

Now say the sentences aloud.

GRAMMAR

1 How many of the sentences in *Sounds* are true for you? Tick (✓) the boxes.

1 ❑   2 ❑   3 ❑   4 ❑   5 ❑

2 Write six sentences about your family or a family that you know well.

1 *My sister play the harp*
2 *My father play the piano*
3 _____
4 _____
5 _____
6 _____

3 Rewrite these sentences in another way. Use the words in brackets.

1 I am his son. (father)
He is my father.

2 You are our sisters. (brothers)
We are your brothers

3 She is my cousin. (cousin)
I am her cousin

4 We are their boyfriends. (girlfriends)
They are our girlfriends

5 He is her husband. (wife)
She is his wife

6 They are your uncles. (nephews)
We are their nephews

4 Is the first word in these phrases singular (S) or plural (P)?

1 church's seats — S
2 cities' names — P
3 girls' sister — P
4 host's flowers — S
5 men's jobs — P
6 people's homes — P
7 Philippa's address — S
8 woman's husband — S

5 Underline the correct word.

1 I've got a brother and a sister. My nieces are my *sister's/sisters'* children.
2 Joanna is my *brother's/brothers'* wife.
3 What are your *cousin's/cousins'* names?
4 This is a photo of my *friend's/friends'* house. They live in the country.
5 Philip and Ray are my *sister's/sisters'* husbands.
6 Is this your *nephew's/nephews'* girlfriend?

LISTENING

1 Look at the photo. Write six questions about this person.

1 Who _____ ☐
2 Where _____ ☐
3 How _____ ☐
4 What _____ ☐
5 _____ ☐
6 _____ ☐

2 🔊 Listen to a man and a woman talking about the person in the photo. Does the man ask any of the questions you wrote in activity 1? Put a tick (✓) by the numbers of the questions that he asks.

3 Write the answers to the questions you wrote in activity 1, where possible.

1 _____
2 _____
3 _____
4 _____
5 _____
6 _____

🔊 Now listen again and check.

SPEAKING

1 You are going to talk about someone you know well. Think about what you're going to say. Use your questions in *Listening* activity 1 to help you.

2 Record yourself on a blank cassette.

SOUNDS

▭▭ Listen to these questions. Put a tick (✓) if you think the speaker sounds interested.

1 Has it got many restaurants?

2 Is it an expensive city to live in?

3 How good is the public transport?

4 How many museums are there?

5 Has it got a sports stadium?

6 Is there a university?

Say the sentences aloud. Try to sound interested.

VOCABULARY AND READING

1 Write the names of features or facilities near your home.

cinema _____ _____

park _____ _____

_____ _____

_____ _____

_____ _____

_____ _____

2 Read the *Resort fact files*. Complete these sentences with the names of the resorts.

1 *Kalo Chorio* hasn't got a doctor.

2 *Aghios Nikolaos* has got more than one chemist.

3 *Kalo Chorio* hasn't got a bank.

4 *Kalo Chorio* has got one disco.

3 Answer these questions.

1 How many banks has Aghios Nikolaos got?

It's got three.

2 Have the resorts got any supermarkets?

Yes, it has

3 How many post offices has Kalo Chorio got?

It's got one.

4 Which resort hasn't got a market?

It's Kalo Chorio

CRETE

Resort fact file: AGHIOS NIKOLAOS

General character: Modern and lively.
Highlights: Dinner by the lake; dancing till dawn.
Negatives: No peace and quiet!
Beach: One small sandy beach and two stony beaches.
Discos: Twelve.
Bars: Too many to count!
Eating out: A good variety: excellent food.
Supermarkets: Two. Lots of small food stores and a weekly market.
Chemist: About twelve.
Doctor: Dr Kounelakis, October Street.
Exchange facilities: Three banks on the main square.
Post office: Yes, near the square.
Transport: Regular bus service all over the island.

Resort fact file: KALO CHORIO

General character: Relaxed atmosphere with very friendly locals.
Highlights: The finest beaches in the area!
Negatives: Not many watersports out of the summer season.
Beach: Two beaches to choose from, one of which is fine golden sand.
Discos: One.
Bars: Quite a lot of bars. Try Pinnochio's for a friendly atmosphere and a picnic or barbecue.
Eating out: About ten restaurants with good food and great service.
Supermarkets: Yes.
Chemist: One.
Doctor: Dr Kounelakis, October Street, Aghios Nikolaos (11 km).
Exchange facilities: Post office; hotels and many shops will change money.
Post office: One.
Transport: Taxis; bus service every half hour.

GRAMMAR

1 Write questions about the two holiday resorts for these answers.

1 *How many supermarkets has Aghios Nikolaos got?*

It's got two, and it's got lots of small food shops.

2 *Does Aghios Nikolaos has a Doctor*

Yes, it has. He lives in October Street.

3 *Does Kalo Chorio has a Beach*

Yes, it has. It's got the best ones in the area.

4 *Does Aghios Nikolaos have a post office*

Yes, it's got one near the square.

2 Complete the sentences with *have* or *has*.

1 She *has* three sisters.
2 We *have* got a park near our house.
3 Madrid *has* got a very famous museum – the Prado.
4 I *have* my own car.
5 They *have* got a studio near the centre of the city.
6 It *has* got a terrace on the roof.

3 Rewrite the sentences in activity 2 with the contracted form of the verbs if possible.

_____
_____
_____
_____
_____
_____

WRITING

1 Join the two ideas. Write sentences with *and* or *but* about a city you know.

1 architecture is interesting, streets are dirty
The architecture is interesting but the streets are dirty.

2 public transport is safe, public transport is clean

3 excellent restaurants, expensive restaurants

4 crowded, dangerous

5 a couple of parks, no swimming pools

6 shops are cheap, shops are crowded

2 Read Diane's postcard and decide where these phrases can go.

1 and the weather's great
2 but it's very relaxing
3 and having a drink and a barbecue
4 but there are quite a lot of bars
5 and it's only five minutes' walk from the shops and tavernas
6 but I'm just red

Here we are in Kalo Chorio. We're having a lovely time. Martin's nice and brown. We've got a nice apartment. There isn't much to do here. There's only one disco. We're sitting in Pinnochio's. Hope all's well at home. See you soon.
Love, Diane x x

3 Imagine you are in your dream holiday resort. Write someone a postcard.

11 | How ambitious are you?

GRAMMAR

1 Match the two parts of the sentence. Write the numbers in the boxes.

1 I'd like to buy a horse
2 I'd like to work abroad
3 I'd like to run a marathon
4 I'd like to lose some weight ->le poit
5 I'd like to try parachuting
6 I'd like to change my job

[6] ... so I'm going to look at the ads in tomorrow's paper. ✓

[4] ... so I'm not going to eat junk food. ✓ *inca-brac*

[2] ... so I'm going to teach myself a foreign language. ✓

[3] ... so I'm going to do some serious training. ✓

[1] ... so I'm going to start saving some money. ✓

[5] ... so I'm going to phone the airport and see if you can have lessons. ✓

2 Rewrite the sentences in activity 1 with *because*.

1 *I'm going to start saving some money because I'd like to buy a horse.* ✓
2 *I'm going to teach myself a foreign language because I'd like to work abroad* ✓
3 *I'm going to do some serious training because I'd like to run a marathon.* ✓
4 *I'm not going to eat junk food because I'd like to lose some weight.* ✓
5 *I'm going to phone the airport and see if you can have lessons because I'd like to try parachuting.* ✓
6 *I'm going to look at the ads in tomorrow's paper because I'd like to change my job* ✓

3 Write sentences about your own ambitions and plans. Use these verbs.

1 be   2 buy   3 get
4 go   5 have   6 meet

1 _____
2 _____
3 _____
4 _____
5 _____
6 _____

4 Write questions with *going to*.

1 I've got a car but I can't drive.
   Are you going to learn to drive?

2 I like getting up late at the weekend.
   Are you going to get up late this weekend?

3 I've got a day off work tomorrow.
   What are you going to do?

4 I usually run home after work but it's raining today.
   Are you going to catch a bus

5 There's a great film on TV this evening.
   Are you going to watch it

6 My flat's in a mess.
   Are you going to be in a mess?

5 Complete the sentences with one of the verb forms in brackets.

1 I like *going* to the cinema. (going/to go)
2 My sister would like *to have* a dog. (having/to have)
3 I enjoyed *being* on holiday. (being/to be)
4 A couple of days ago my father decided *to change* his car. (changing/to change)
5 My parents want *to live* in the country. (living/to live)
6 I don't mind *doing* my homework. (doing/to do)

6 Write true sentences about yourself or people you know. Use the six other verb forms in activity 5.

1 _____
2 _____
3 _____
4 _____
5 _____
6 _____

WRITING

1 Read what Natasha says about her plans and ambitions. Why is she going to look after some children this summer?

_____
_____

> *I'm in my last year at school. I'd quite like to work abroad, but first I need to get some qualifications. I've got a place at Leeds University and I'm going to do an economics degree. Languages are important in the business world, so this summer I'm going to look after some children in Germany. I'd like to improve my French too.*

2 Why are you learning English? Do you have any plans and ambitions? Write about them.

_____
_____
_____
_____
_____
_____

SOUNDS

1 Look at what Natasha says about her plans and ambitions in *Writing* activity 1. Listen and underline the stressed words.

Now read the passage aloud.

2 Read aloud the sentences you wrote in *Writing* activity 2.

LISTENING

1 Listen to four people talking about their ambitions and plans. Write the number of the speaker by the correct picture.

2 Write down the speakers' ambitions and plans.

Speaker 1 _____
_____
Speaker 2 _____
_____
Speaker 3 _____
_____
Speaker 4 _____
_____

Now listen again and check.

SPEAKING

1 You are going to talk about your plans and ambitions. Think about what you're going to say.

2 Record yourself on a blank cassette.

23

READING

1 Read the passage below and choose the best title. (You can use a dictionary if you like.)

1 English in the future

2 Language teaching in British schools

3 Are the British bad at languages?

4 The British school system

Since 1989, all state schools in Britain offer nine subjects: art, English, geography, history, mathematics, music, physical education, science and technology. At secondary schools, pupils from the age of 11 also learn a modern foreign language.

At the moment, most pupils choose French, and European languages will probably be the most popular ones in the near future. But Britain has a high immigrant population and it's possible that schools will also offer Urdu, Gujerati or any one of the Asian languages spoken by the ethnic groups. However, the British government does not participate in the European *Lingua* programme, which requires all member states to offer two European languages in their schools. In general, people don't think foreign languages are important, and few people speak a foreign language fluently.

But many people have strong opinions about how schools should teach their own language – English. Some think schoolchildren should speak with the same standard accent and stress the importance of grammar. Others feel that regional accents and dialects are just as important, and part of someone's cultural identity. It is clear, however, that the type of language a person uses shows a lot about their education and background. Teaching English as a mother tongue will probably remain more important than learning a foreign language.

2 Look at the sentences which are underlined. Are they also true for your country?

3 Is there any other information about Britain which is also true for your country?

4 Read the passage again and write down any facts about the British education system. Write down similar facts about the education system in your country.

Britain

- At secondary schools, pupils from the age of 11 also learn a modern foreign language.
- most pupils choose French or Asian languages
- some think schoolchildren should speak with the same standard accent
- nine subjects

Your country

- this is the same in French
- most pupils choose English but few pupils choose German
- In France, the people has got different accent but it is all the same

SOUNDS

Underline the /ɒ/ sounds and circle the /əʊ/ sounds.

1 She won't take her shoes off.

2 It'll be there at six o'clock.

3 We don't want a large meal.

4 He likes discos a lot.

5 I'll stop smoking soon.

6 I go to school on my bike.

🔊 Listen and check. Say the sentences aloud.

VOCABULARY

1 Write down the school subjects you need to study to do the following jobs.

accountant _math , ict_

doctor _science , biologie_

engineer _math , physics, ict ,_

journalist _languages, ict_

politician _economics , humains science , languages_

secretary _ict , languages_

2 Write the subjects you do or did at school in the columns below.

I like(d)	I dislike(d)
_____	_____
_____	_____
_____	_____
_____	_____
_____	_____
_____	_____

3 Write the names of six people you know. Write their jobs next to their names.

names	jobs
_____	_____
_____	_____
_____	_____
_____	_____
_____	_____
_____	_____

4 In which of these situations do you hear or see English? Tick (✓) the boxes.

at school ❑   at work ❑   at home ❑

on holiday ❑   with friends ❑   in shops ❑

in banks ❑   at the airport ❑

at the railway station ❑

GRAMMAR

1 Add *'ll* to these sentences where possible.

1 I' _'ll_ get a better job if I speak English.

2 She ___ has salad for lunch. ✓

3 They _'ll_ give up wine with their meals. ✓

4 I _'ll_ call you when I get there. ✓

5 He ___ does a lot around the house. ✓

6 It ___ rains in the afternoon. ✓

2 Make predictions about people's futures. Use four words – one from each box – for each prediction.

become	his	and	business
decide	married	foreign	famous
get	rich	in	friends
learn	some	leave	home
make	to	new	language
start	another	own	September

1 I think Susan _will become rich and famous_

2 I think Peter _will decide married in September_

3 I think Katie _will make some new friends_

4 I think William _will learn another foreign language_ ✓

5 I think Joanna _will get his own business_

6 I think Eddie _will start to leave home_

3 Write some predictions for people you know for the next year.

1 _____

2 _____

3 _____

4 _____

5 _____

6 _____

WRITING

How many of your friends or family speak English? (Don't include people in your class.) Write a paragraph about one of these people. Say when, where and why this person speaks English.

13 | Foreign travels

GRAMMAR

1 Underline the best verb form.

1 'I need some shopping.' *I'm going to/I'll* do it for you, if you like.'

2 *She's going to/She will* take her camera because her hobby is photography. ✓

3 You look tired. *I'm going to/I'll* drive you home.' ✓

4 'Where *are you going to/will you* spend your holidays?' ✓ '*We're going to /We'll* stay with friends in Cornwall. We've just arranged the dates.' ✗

5 Let's have lunch together. *We're going to/We'll* try that new restaurant in the High Street. ✓

6 Duncan is coming round at eight this evening. *He's going to/He'll* probably be late. ✗

2 Which comes first? Put these expressions in order by numbering the boxes.

6	in a year's time ✓
4	in three days' time ✓
5	next month ✓
3	the day after tomorrow ✓
1	this evening
2	tomorrow morning ✓

3 Use the expressions in activity 2 and talk about things which are going to happen and things that will perhaps or probably happen.

1 I am going to do my Homework this evening ✓

2 I would like to get up late tomorrow morning. ✓

3 I am going to work hard the day after tomorrow ✓

4 I would like to go swimming with my parents in three days' time ✓

5 I would like to be happy next month ✓

6 I am not going to have a dog in a year's time ✓

4 Complete John's sentences with *'ll* or *won't* to talk about his positive decisions for changes for the future.

1 I 'll eat less fast food.

2 I won't be late for lessons. ✓

3 I won't be so pessimistic. ✓

4 I 'll walk to work and get some exercise. ✓

5 I won't go to bed so late. ✓

6 I 'll stop watching so much television. ✓

5 Rewrite the sentences in activity 4. Use *'ll* where you used *won't* and *won't* where you used *'ll* to express the same ideas.

1 I won't eat so much fast food.

2 I'll be on time for lessons

3 I'll be optimistic ✓

4 I won't try to work

5 I'll go to bed early ✓

6 I won't watch so much television

6 Veronique wants to be a better student. Complete her decisions with *'ll* or *won't*.

1 I won't forget to do my homework.

2 I 'll learn six new words every day. ✓

3 I won't arrive late for my lessons. ✓

4 I 'll buy an English newspaper. ✓

5 I 'll speak English when we work in pairs. ✓

6 I won't be so lazy. ✓

7 Make some decisions about your own language learning.

1 I'll speak more at the teachers.

2 I'll read some English books.

3 I won't be so shy (or bashful)

4 I'll learn more hard

5 I won't forget a lesson

6 I'll do some extra works

8 Complete these sentences with *going to* or *will*.

1 You use *going to* for plans.

2 You use _will_ for decisions you take at the moment of speaking. ✓

3 You use _going to_ to talk about things which are arranged or sure to happen. ✓

4 You use _will_ to make a prediction. ✓

9 Match these sentences with the rules in activity 8. Number the boxes.

☐2 I'll make you a drink.

☐1 I'm going to visit my grandparents at the weekend.

☐4 I'll go there on Saturday morning.

☐3 My dad's going to phone at six o'clock.

SOUNDS

Say these words. Which of these words have the sound /æ/? Underline the sound.

aspirin backpack camera currency food
guide book handbag map medical kit
passport penknife razor scissors sleeping bag
suitcase tent toothbrush toothpaste
traveller's cheques walkman wallet watch

Now listen and check.

VOCABULARY AND LISTENING

1 Look at the picture and write a list of the things Sheila is going to take with her on holiday to Brazil.

_____
_____
_____
_____
_____
_____

2 Listen to Sheila talking about the things she's going to take on holiday. Which two things in the picture isn't she going to take?

3 Why she isn't going to take the two things?

Listen again and check.

READING

1 You are going to read about teatime in England. Do you have anything to eat or drink at around 4pm in your country?

2 Read *Teatime* and decide if you would like to have a typical English tea. If so, which tea shop in Oxford would you like to visit?

3 Look at the map and find the tea shops mentioned. Put the number of the tea shop by its name in the text.

4 Read *Teatime* again and note down things you eat and drink at a typical English tea. (You can use a dictionary to find out what they mean.)

Teatime

> *'I'll have cucumber sandwiches, crumpets, scones with jam and cream, a slice of Dundee cake and a pot of Earl Grey tea, please. But no sugar – I'm on a diet.'*
> **Overheard in an English tea shop.**

In England, tea is not just a drink, it's a meal as well. In some homes it's the last meal of the day, although for many people it's a light meal between lunch and dinner. But for most people at work there isn't time to stop for a proper tea. Most tea shops serve tea between three and five in the afternoon, although the most popular time is around four o'clock. So an English tea is ideal for people on holiday who have the time to enjoy it. And why not try a typical English tea in that most English of cities, Oxford? After a long day visiting the romantic city of dreaming spires, here are some tea shops to visit.

The Randolph Hotel in Beaumont Street is in the very centre of the city. It's a fine Victorian building and offers the best traditions of English teatime, with piano music in the background.

The Nosebag in St Michael's Street is popular with the university students. Excellent homemade cakes and a good choice of teas are available but go early as there are always long queues.

The Wykeham Coffee Shop in Holywell Street is a small tea shop opposite New College, and is popular with tourists and students.

George's Cafe in the Covered Market is excellent value and a favourite with market workers and shoppers.

Queen's Lane Coffee House on the High Street is one of the oldest restaurants in Oxford, and despite its name, serves excellent teas as well – and a choice of newspapers.

But the best teatime suggestion is to make friends with some poor students and offer to buy them tea – served in their own college common room!

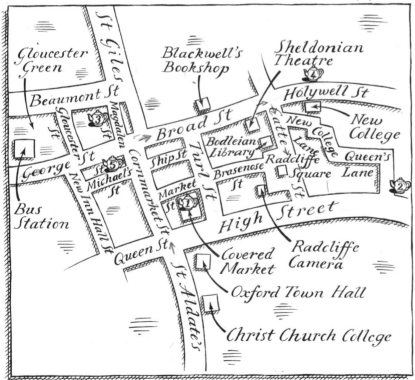

VOCABULARY

1 Look at the map again. Can you see any of these town features on the map? What are the places called?

bank   bus station   college   hospital   law courts
library   parliament   pub   railway station
river   town hall

2 Does your town have any of the features in activity 1? What are they called?

_____
_____
_____
_____
_____
_____

SOUNDS

Say these words aloud. Underline the three words in each group with the same-sounding ending.

1 colleges   houses   offices   theatres
2 factories   libraries   pubs   universities
3 hotels   hospitals   stations   supermarkets
4 banks   squares   restaurants   shops

🖭 Now listen and check.

GRAMMAR

1 Look at the map and describe where the following places are.

The Town Hall
is behind the Christ Church College
New College
is on the corner of Holywell St and New College
The Bodleian Library   Queen's
is in front of the Sheldonian Theatre
The Covered Market
is between Market St and High St.

2 You're standing in Radcliffe Square and you want to go to the Randolph Hotel. Use the map to complete this guided tour.

The Radcliffe Camera is (1) *in front* of you. Walk across the square and turn (2) left into the Bodleian Library, which is (3) next to the Sheldonian Theatre. Cross (4) over Broad Street and visit Blackwell's Bookshop. Then walk to the (5) end of Broad Street and turn (6) right into St Giles. At the crossroads, turn (7) left into Beaumont Street. The Randolph Hotel is (8) on the (9) left .

3 Write directions for the following.

1 from Christ Church College to the Sheldonian Theatre

The Christ Church College is in front of you. Walk to the end of St Aldate's St. You go in the Cornmarket St, walk to the end of this St. Turn right in Broad St. The Sheldonian Theatre is on the right before the Catte St.

2 from the Bus Station to Radcliffe Square

The Bus Station is in front of you. Walk to the end of George St and Broad St. Then turn right in Catte St. The Radcliffe Square is on the corner of Catte and Brasenose

3 from New College to the Covered Market

The New College is on your left side. Go straight from Holywell St to Broad St. Up to Turl St. Take the second turning on the right. Covered Market is straight on your left side.

WRITING

An English friend is staying at the Randolph Hotel. Write a message to your friend and suggest you have tea in Queen's Lane Coffee House. Give him/her directions and suggest a time to meet.

15 | *An apple a day*

VOCABULARY

1 Circle the odd-word-out in each line.

1 beer   coffee   lettuce   water
2 beef   chicken   lamb   fish
3 apple   carrot   onion   potato
4 banana   orange   peach   vegetable
5 butter   cheese   egg   juice
6 bread   milk   tea   wine

2 Which words from activity 1 can you put with these words or expressions?

1 An   *apple, onion, orange, egg*

2 A glass of _____

_____

3 A packet of _____

_____

4 A cup of _____

_____

5 A slice of _____

_____

6 A piece of _____

_____

3 Write down the name of the food that each picture shows. Use *a/an* or *some* and one other word.

1 *an egg*   *some eggs*   *some egg*

2 *some* _____   *some* _____   *a* _____

3 *some* _____   *a* _____   *some* _____

4 *a* _____   *some* _____   *some* _____

SOUNDS

🔲 Listen to the sentences and notice how *a* is pronounced /ə/, *an* is pronounced /ən/, and *of* is pronounced /əv/.

1 I'd like an orange.
2 I'd like a peach, please.
3 There are seven pieces of cake.
4 I need a glass of water.
5 She needs a cup of tea.
6 I like bread and butter.

Now say the sentences aloud.

LISTENING

1 🔲 Listen to two people planning dinner. Put a tick (✓) by the things they need to cook dinner.

☐ cabbage   ☐ lemon juice   ☐ egg
☐ onion   ☐ orange   ☐ bread
☐ chicken   ☐ cream   ☐ rice
☐ sugar   ☐ tomatoes   ☐ wine
☐ olive oil   ☐ peaches   ☐ mushrooms

2 Which of the things do they need to buy? Write their shopping list.

an onion
_____
_____
_____
_____

🔲 Listen again and check.

3 Would you like to have dinner with the two people? Why (not)?

_____
_____

30

GRAMMAR

1 Underline the best verb form.

1 *I like/I'd like* a slice of chicken.

2 There's Peter! *He's having/He has* a cup of tea. ✓

3 *I'm eating/I eat* a cheese sandwich. ✓

4 *I like/I'd like* all types of fruit. ✓

5 *Do you/Would you* like some salad? ✓

6 *We're drinking/We drink* wine sometimes. ✓

2 Write questions with *How much* and *How many*.

1 Shall we get some tomatoes?

How many shall we get?

2 We need some cheese.

How much do need we?

3 I'd like some oranges.

How many would you like? ✓

4 I've got some coffee.

How much have you got? ✓

5 Shall we get some bread?

How much shall we get? ✓

3 Complete these sentences with *some* and *any*.
Then write another sentence with *(don't) need*
or *have(n't) got*.

1 We've got *some* apples.

We don't need any apples.

2 I don't need _any_ lettuce. ✓

I have got any lettuce
some

3 He hasn't got _any_ eggs. ✓

He need some eggs

4 She needs _some_ fish. ✓

She hasen't got any fish ✓

5 I've got _some_ milk. ✓

I don't need any milk ✓

READING

1 You're going to read a passage about shopping
for food. Do you like or dislike shopping for
food? Can you say why?

2 Read *A nation of shoppers*. What do you think
is the most interesting piece of information?

A NATION OF SHOPPERS

'You are what you eat.'
German proverb

*'You eat what you shop.
You are what you shop.'*
Graffito in London

**The food you eat, as well as where and when
you buy it, says a lot about you as an
individual and about your cultural background.
We talked to some British shoppers about
their shopping habits.**

**Who does the shopping for food and drink in
your house?**
▨ 'We both do the shopping. I drive her to the
supermarket on Saturday mornings, and then I
wait for her to do the shopping.' *Paul, Gateshead*

What items of food do you buy every week?
▨ 'Potatoes, meat, cheese, butter, green
vegetables, pasta, eggs, tea, and some fruit,
maybe some apples and bananas. I also buy a lot
of frozen food.' *Tim, Chiswick*

What items of food do you buy every day?
▨ 'Well, I don't always go shopping every day.
There's nothing really fresh which I need, except
for milk. And the milkman delivers that to my front
door every morning.' *Mary, Derry*

Do you always buy the same things?
▨ 'Yes, for the family. I sometimes buy different
things if we're going to have guests. I love to try
recipes for Italian food.' *Fiona, Inverness*

How often do you go shopping for food?
▨ 'Twice a month. We go to the hypermarket on
the edge of town and buy a lot. We live busy lives
and we don't have time to shop very often.'
Henry, Nottingham

How often do you go shopping in a market?
▨ 'I love to go to the market at the weekend. I
take the children and we buy lots of fresh
vegetables and meat. I wish I could do it every
day.' *Frances, Abergavenny*

3 Think about your answers to the questions in
the passage. Are your answers to the questions
the same as the answers in the text?

4 Do you think most people in your country have
the same answers?

VOCABULARY

Look at a calendar or diary for this year. Complete the list below with the day of the week or the month.

1 Monday 1st *August*
2 _____ 26th February
3 Thursday 27th _____
4 _____ 5th December
5 Friday 11th _____
6 _____ 4th January
7 Tuesday 10th _____
8 _____ 21st August
9 Wednesday 22nd _____
10 _____ 17th June
11 Saturday 3rd _____
12 _____ 13th October

GRAMMAR

1 Write these expressions in three columns.

August   Friday evening   half past five   night
1994   2nd November   Sunday 16th May
three o'clock   the morning   the weekend
Wednesday   winter

at	in	on
half past five	August ✓	Friday evening ✓
night ✓	1994 ✓	2nd November ✓
three o'clock ✓	the morning ✓	Sunday 16th May ✓
the weekend ✓	winter ✓	Wednesday ✓

2 Choose six expressions from activity 1 and write sentences about yourself.

1 _____
2 _____
3 _____
4 _____
5 _____
6 _____

3 Put these words in order and make invitations.

1 go Saturday let's on shopping morning
Let's go shopping on Saturday morning.

2 about to in afternoon how going the theatre the
How about going to the theatre in the afternoon? ✓

3 we dinner evening don't in why have the
Why don't we have dinner in the evening? ✓

4 French you try to the like restaurant would new
Would you like to try the new French restaurant? ✓

4 Accept or refuse the invitations in activity 3.

1 I'm sorry but I can't because my sister has for birthday ✓
2 I'd love to go to the theater.
3 I'm busy. I have promess my brother to go to the park
4 I'd love to try the new French restaurant ✓

5 Think of four things you are going to do soon. Invite people to do these things with you.

1 _____
2 _____
3 _____
4 _____

6 Complete these sentences with *I* or *I'd*.

1 *I* like watching TV in bed.
2 _____ love to join you at the concert.
3 _____ like Dustin Hoffman's films.
4 _____ don't like sandwiches very much.
5 _____ like to go to the ballet with you.
6 _____ like to work in New York City.

SOUNDS

1 🔊 Listen to these phrases. Notice how you don't always hear the *'d* sound.

1	I like	I'd like
2	we love	we'd love
3	they like	they'd like
4	I love	I'd love
5	we like	we'd like
6	they love	they'd love

2 Now look at the phrases in context and underline the correct verb form.

1 *I like/I'd like* to go to the cinema.
2 *We love/We'd love* to visit China.
3 *They like/They'd like* playing golf.
4 *I love/I'd love* to live near a beach.
5 *We like/We'd like* watching old films.
6 *They love/They'd love* doing crosswords.

🔊 Listen and check. Say the sentences aloud.

READING

1 Look at the advertisements. What are the four types of entertainment?

1 _____ 3 _____
2 _____ 4 _____

1
Showcase Cinema
Jurassic Park
Daily: 2.10, 5.10, 8.10
Tickets: £4 (£2 with student's card)
Advance booking
ACCESS/VISA

2
Arts Centre
Guys and Dolls
Frank Loessor's hit musical
Mon–Sat 7.30, ends Saturday
Tickets £7–£9

3
THE GALLERY
Horse art
Sculpture and pictures.
Wed–Sun 1 pm–5.30 pm
until August 22nd.
Free.

4
Last two performances today!
THE PLAYHOUSE PRESENTS
KIROV BALLET
Swan Lake
2.30 pm and 8.00 pm
Tickets from £8

2 Are these sentences true or false? Put T or F in the boxes.

1 Tickets for the musical cost £4. `F`
2 The exhibition is open on Fridays. ☐
3 The gallery is open in the mornings. ☐
4 There are no tickets left for the Kirov Ballet. ☐
5 The film's on three times every day. ☐
6 Tickets for the musical are all the same price. ☐
7 There's a play at one of the places. ☐
8 The film lasts four hours. ☐

3 Look at the entertainments in activity 1. Which would you most like to go to? Why?

_____
_____

LISTENING

1 🔊 Louise is looking at the newspaper ads in *Reading* and telling her friend what's on. Listen and find out where they decide to go.

2 Which of these questions does Louise's friend ask? Tick (✓) the correct boxes.

☐ What's on?
☐ Where's it on?
☐ When's it on?
☐ Who's in it?
☐ Who's playing?
☐ What time does it start?
☐ How much does it cost?

🔊 Listen again and check.

3 Write notes with details about the choice of entertainment, place, time, and meeting place.

_____
_____
_____

4 Invite a friend to one of the other entertainments. Write your conversation. Use the tapescript of Louise's conversation to help you.

VOCABULARY

1 Write these words in the two lists below.

attractive   calm   confident   *confiant*   good-looking
intelligent   nervous   pretty   *mince* slim   quiet   *goli*
grand tall   *pensif* thoughtful   *laid* ugly

appearance	character
attractive	calm
good-looking ✓	intelligent ✓
pretty ✓	nervous ✓
slim ✓	quiet ✓
tall ✓	confident ✓
ugly ✓	thoughtful ✓

2 Look at the pictures. Match these descriptions with three of the people. Write the numbers in the boxes.

1   **2**   **3**   **4**   **5**

She's medium-height and middle-aged. She's got dark, curly hair and glasses.   [3] ✓

He's got a moustache and a beard. He's bald. He's short and fat.   [2] ✓

She's got long hair. It's fair and straight. She's tall and thin. *neat*   [1] ✓

3 Complete this description of one of the other two people.

He's got _dark_ hair and a _beard_. He's _tall_ , _slim_ and good-looking. ✓

4 Write a description of the fifth person.

She is old and short. She wears glasses, ...

SOUNDS

Put the words in order and make questions. Then underline the stressed words.

a   hair   long's   her   how
How long's her hair?

b   she   how   is   old
How old is she? ✓

c   who's   like   she
Who's she like? ✓

d   she   like   what   look   does
What does she look like? ✓

e   look   who   like   does   he
Who does he look like? ✓

f   like   he   what's
What's he like? ✓

g   is   how   he   tall
How tall is he? ✓

h   hair   what   his   colour's
What colour's his hair? ✓

🔊 Now listen and check. Say the sentences aloud.

LISTENING

1 🔲 Listen to a man answering questions about a friend. In which order does he answer the first four questions in *Sounds*? Write the letter of the question in the correct box.

1 ☐ 2 ☐ 3 ☐ 4 ☐

2 🔲 Listen to a woman answering questions about a friend. In which order does she answer the last four questions in *Sounds*? Write the letter of the question in the correct box.

1 ☐ 2 ☐ 3 ☐ 4 ☐

GRAMMAR

1 Complete these sentences about yourself with *quite, (not) very* and *(not) really*.

1 I'm ___quite___ attractive.
2 I'm ___not very___ old.
3 I'm ___not very___ slim.
4 I'm ___quite___ tall.
5 I'm ___very, very___ kind.
6 I'm ___very___ shy.
7 I'm ___very___ patient.
8 I'm ___not very___ lazy.

2 Write questions for these answers.

1 *How tall is she?*
   About one metre fifty.
2 How old is he?
   Middle-aged, I'd say.
3 How long's her hair?
   It's not very long at all.
4 Who's she like?
   Like a typical teenager.
5 What does she look like?
   She's got her mother's eyes but her father's hair.
6 What colour's his hair?
   It was brown but he's bald now.
7 Who does she look like?
   She's like her father – rather sensitive and shy.
8 What's he like?
   He's quite good-looking, with a beard and glasses.

3 Choose a man and a woman you know well. Answer these questions.

1 Has the man got a beard?
   ___no___
2 Is he middle-aged?
   ___yes___
3 Has the woman got blonde hair?
   ___no___
4 Does she wear glasses?
   ___no___
5 Is the man confident?
   ___yes___
6 Does he look like his parents?
   ___yes___
7 Is the woman like her mother?
   ___no___
8 Does she look like a businesswoman?
   ___no___

WRITING

Write a description of either the man or the woman you chose in *Grammar* activity 3. Use *quite, very* and *really* where possible.

_____
_____
_____
_____
_____
_____
_____

SPEAKING

1 You are going to talk about your appearance and character. Think about what you're going to say. Use the questions in *Sounds* to help you.

2 Record yourself on a blank cassette.

How to be an American

Despite all our different origins, there is a style that marks North Americans. We have shy people and bold, talkative and quiet, and yet you cannot mistake the quality of 'Americanness'.

★ Nearly everyone agrees that we are friendly. It shows that we think everyone is equal and has rights. But being friendly is different from being friends. True friends are as difficult to make here as everywhere else.

★ Americans do not think it necessary to hide their emotions. Sometimes we seem to exaggerate them. We show happiness with big smiles, gestures and exaggerated statements. Unlike many Asians, Americans smile only around good news or happy feelings. An American smiles often, but not when embarrassed or confused. Bad news does not come with a smile.

★ Because we don't touch each other very much, Latin Americans find us cold. We have a strong sense of private space. We stand at least an arm's length apart, and are made uncomfortable by people who stand closer.

★ Because it is important to be assertive, Americans speak fairly loudly. Foreigners sometimes mistake the loudness for anger, especially because it is more acceptable to show anger than in other cultures.

★ Visitors usually find Americans very polite, because we use *please* and *thank you* very often and because of the way we are friendly to strangers. We are polite to waiters and garage attendants as well as doctors and senators.

★ We are very informal. The forms of our language do not change when we talk to a superior, as they do in many languages. People dress casually and use first names most of the time. But it is still easy for a foreigner to make mistakes and be too informal in the wrong circumstances.

Adapted from *Culture Shock USA*
by Esther Wanning

READING

1 The article *How to be an American* is about the national characteristics of North Americans. Read it and decide if the writer is an American or not.

2 Write down adjectives the writer of the article might use to describe the characteristics of Americans.

_____

_____

_____

_____

Do you agree with the writer?

3 How similar are your national or regional characteristics to those of North Americans? Write a few adjectives describing your national or regional characteristics.

_____

_____

_____

_____

4 Do you think people from other countries would describe your national or regional character in the same way?

SOUNDS

Listen to these sentences. Notice the stress and intonation in sentences with lists.

1 She's imaginative.
2 She's imaginative, clever and confident.
3 He's serious.
4 He's serious, thoughtful and shy.
5 She's nervous.
6 She's nervous, sensitive and kind.

Now say the sentences aloud.

VOCABULARY AND LISTENING

1 Choose six words from the following list to describe your ideal partner. Underline the words.

calm careful clever confident friendly
honest imaginative intelligent interesting
kind lazy nervous nice optimistic tidy
patient pessimistic polite quiet reliable
sensible sensitive serious shy thoughtful

2 🔲 Listen to three people describing their partners. Put a tick by the words in activity 1 that they use to describe them. Then write the words in two columns.

his/her partner...

is	isn't
Speaker 1	
_____	_____
_____	_____
Speaker 2	
_____	_____
_____	_____
Speaker 3	
_____	_____
_____	_____

3 Do the speakers admire these qualities in their partners? Tick (✓) the qualities they admire.

🔲 Listen again and check.

GRAMMAR

1 Complete the chart.

	adjective	comparative	superlative
1	*bad*	worse	*worst*
2	tall	taller	tallest
3	close	closer	closest
4	good	better	best
5	friendly	friendlier	friendliest
6	kind	kinder	kindest
7	nervous *(gentil)*	more nervous	most nervous
8	nice	nicer	nicest
9	sensitive	more sensitive	most sensitive
10	smart *(petit)*	smarter	smartest

2 Complete these sentences with words from the chart in activity 1. The sentence numbers correspond to the numbers in the chart.

1 I'm really _____*bad*_____ at maths.
2 She's _____taller_____ than me.
3 My brother and I aren't very _____close_____.
4 My sister's my _____best_____ friend.
5 Is she _____more friendly_____ than her sister?
6 My boss's not very _____kind_____.
7 I don't think I'm the _____nicest_____ person in the class.
8 She's the _____nicest_____ person I know.
9 People often become _____more sensitive_____ when they get older.
10 That's a very _____smart_____ suit.

3 You are going to read about three friends. Use superlative forms of these adjectives and write questions beginning *Who is the ... ?*

1 nice   2 lazy   3 young   4 short
5 pessimistic   6 good-looking

1 *Who is the nicest?*
2 Who is the laziest?
3 Who is the youngest?
4 Who is the shortest?
5 Who is the most pessimistic?
6 Who is the most good-looking?

4 Read about the three friends and answer the questions you wrote in activity 3.

1 Bob   2 Carl   3 Bob
4 Alan   5 Bob   6 Alan

1 Alan isn't as nice as Bob but he's nicer than Carl.
2 Carl's lazier than the others and Bob isn't as lazy as Alan.
3 Carl's the oldest. Bob isn't as old as Alan.
4 Bob's as tall as Carl but Alan isn't as tall as they are.
5 Carl's a bit more optimistic than Alan. Bob isn't as optimistic as either of them.
6 Bob is really ugly. Carl and Alan aren't very good-looking but Alan is more good-looking than Carl.

VOCABULARY

1 Look at the pictures. Which people are wearing these clothes? Write B (boy), G (girl) M (man) or W (woman) in the boxes. There are four items of clothing that none of the people are wearing.

a blouse ☐   a coat [M]   a dress ☒
a hat ☒   a jacket [W]   jeans ☐
a shirt ☐   shoes ☐   a skirt [W]
socks ☐   a suit ☐   a sweater ☐
a swimsuit ☐   a tie [M]   tights ☐
trainers ☐   a T-shirt ☐

2 What are you wearing now? Write a list.

_____
_____
_____

3 Complete these sentences about your clothes.
1 I've got a black _____.
2 I haven't got a red _____.
3 I've got some brown _____.
4 I haven't got any yellow _____.
5 I've got some white _____.
6 I haven't got any pink _____.
7 I've got a grey _____.
8 I haven't got a green _____.
9 I've got a blue _____.
10 I haven't got an orange_____.

4 Put the words in order and write sentences.
1 wear white always socks I
   I always wear white socks.
2 my sometimes jeans sister wears
   _____
3 tights she black often wears
   _____
4 a her brother hat never wears
   _____
5 my T-shirts often children wear red
   _____
6 wear I formal never clothes
   _____

SOUNDS

1 Match the words with the same vowel sound.
1 orange   ☐ blouse
2 green   ☐ trainers
3 red   [1] socks
4 white   ☐ jeans
5 brown   ☐ tights
6 grey   ☐ sweater

🔊 Now listen and check. Say the words aloud.

2 Look at this true sentence.

The man's wearing a brown suit.

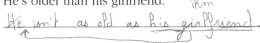 Listen and correct the statements below with the true sentence. Change the stressed word each time.

1 The woman's wearing a brown suit.

2 The man's buying a brown suit.

3 The man's wearing a grey suit.

4 The man's wearing a brown jacket.

GRAMMAR

1 Complete the chart.

adjectives	comparative forms
good	better
great ✓	greater
late ✓	later
big ✓	bigger
lazy ✓	lazier
formal ✓	more formal

2 Think about the comparative forms of these adjectives. Underline the odd-one-out.

1 <u>expensive</u> fine large nice

2 hot <u>old</u> ✓ slim thin

3 <u>bad</u> ✓ easy happy tidy

4 cheap tall <u>thoughtful</u> young ✓

5 casual difficult <u>great</u> ✓ modern

6 cold <u>friendly</u> ✓ small smart

3 Rewrite the sentences three times using *less* and *more*.

1 I'm more untidy than my best friend.

I'm less tidy than my best friend.
My best friend's less untidy than me.
My best friend's more tidy than me.

2 I'm less optimistic than my brother.

I'm more pessimistic than my brother ✓
My brother is more optimistic than me ✓
My brother is less pessimistic than me ✓

3 I'm more impolite than my sister.

I'm less polite than my sister ✓
My sister is less impolite than me ✓
My sister is more polite than me ✓

4 Rewrite these sentences. Use *as.*

1 John's brother's taller than he is.

John isn't as tall as his brother.

2 He's older than his girlfriend.

He isn't as old as his girlfriend.

3 Carl's more popular than Bob.

Carl isn't as popular as Bob.

4 He's more casual than his father.

He isn't as casual as his father. ✓

5 His father's more pessimistic than he is.

His father isn't as pessimistic as he.

6 He's thinner than his mother.

He isn't as thin as his mother

5 Write sentences about people you know. Use these adjectives with *as* or *than.*

1 formal   2 thinner   3 more sensitive

4 short   5 worse   6 straight ✓

1 I'm as formal as Cecile. : formaliste

2 Audrey is thinner than Cecile. : mince

3 Laurie is more sensitive than me ✓ : sensible.

4 Thomas is as short as Jean-Victor ✓ : petit

5 I'm worse than Salomé : mauvais ✓

6 I'm as straight as Cecile. : ✓ honnête

WRITING

Write about clothing in your country for the article *Dressing up* in the Student's Book.

20 | *Memorable journeys*

SOUNDS

Listen to these questions. Put a tick (✓) if you think the speaker sounds polite and friendly.

1 How do you get there? ☐

2 Do you go on your own? ☐

3 What time do you leave home? ☐

4 How long does it take? ☐

5 How far is it? ☐

Now say the questions aloud. Try to sound polite and friendly.

GRAMMAR

1 Answer the questions in *Sounds* about your own journey to school/work.

1 _____

2 _____

3 _____

4 _____

5 _____

2 Look at the pictures. Check you can say all the numbers.

3 Write a question which begins with *How* for each picture.

1 *How fast can you go?*

2 _____

3 _____

4 _____

5 _____

6 _____

Now write the answers.

4 Complete the questions with *How* + adjective/adverb.

1 *How   fast*   can you walk?
   About five kilometres an hour.

2 _____ _____ are you?
   I'm eighteen years old.

3 _____ _____ 's your hair?
   Quite short, really.

4 _____ _____ are you?
   I think I'm about 1m 65.

5 _____ _____ 's your bedroom?
   It's about the size of a large car.

6 _____ _____ were your shoes?
   They were £20 – three years ago.

7 _____ _____ do you go to the cinema?
   Only about once a month.

8 _____ _____ brothers have you got?
   I haven't got any brothers. I'm an only child.

5 Answer the questions in activity 4 for yourself.

1 _____

2 _____

3 _____

4 _____

5 _____

6 _____

7 _____

8 _____

VOCABULARY

1 Complete the sentences with eight of these words or expressions. Use the correct form of the verbs.

arrive   border   cost   desert   distance   drive
driver   gallon   gas station   get   highway   hill
leave   mile   mountain   move   home
passenger   petrol   police patrol   reach   set off
speed limit   take   ticket   truck   turn off

(handwritten annotations: le boeuf/la Rivière; un galon; toute nationale/colline coteau; mille; montagne; atteinte/porté partiri/se mettre; limitation de vitesse; wagon, plate forme); en route)

1 The British equivalent of _highway_ is motorway.

2 We _set off_ early and travelled all day.

3 The _speed limit_ in the United States is 55 mph.

4 The largest _desert_ in the world is the Sahara.

5 One _gallon_ is about the same as 1.6 kilometres.

6 Most _police patrol_ cars in Britain are white.

7 Garages in Britain now sell _petrol_ by the litre. *(handwritten: vendre)*

8 There was one _passenger_ in the back seat. *(handwritten: le dos; siège)*

2 Think about a journey you often make or a journey you made recently. Underline the words and expressions in the list in activity 1 which you can use to talk about this journey.

3 Write other words you can use to talk about this journey.

center, car, piano, walk, pavement, cross.

READING

1 Read part of the first chapter of Anne Mustoe's book *A Bike Ride*, in which she describes a memorable journey. What does *it* in the first line refer to?

_____

_____

A Bike Ride

Why are you doing it? Why did you do it? Everyone asks the same question and I give a different answer every time – adventure, travel, time for thought, solitude, challenge, curiosity, historical research, freedom, fun ... all are true and good enough reasons. But why go round the world? And why on a bicycle?

For many years I was the headmistress of a private school for girls in the south of England. In January 1983, when I was fifty years old, I was on holiday with two friends in India. One day I looked out of the bus window and saw a cyclist, a solitary European man, riding across the Great Thar Desert. And I was filled with envy. I wanted to be out there myself on that road on a bicycle, alone and free, feeling the reality of India, not looking at it through a pane of glass. I was not athletic. I was not young. I wasn't a keen cyclist and I didn't even have a bike. I had no idea how to mend a puncture. I hated camping, picnics and discomfort. In fact, my qualifications for a long cycle ride were very poor. But it was the bicycle which had immediate appeal.

2 Why were Anne's qualifications for a long cycle ride very poor?

_____

_____

_____

3 What, in your opinion, are the advantages and disadvantages of the bicycle for a long journey?

_____

_____

4 Why do you think Anne's journey was memorable?

_____

_____

SPEAKING

1 You are going to talk about a journey you often make or a journey you made recently. Think about what you're going to say.

2 Record yourself on a blank cassette.

41

SOUNDS

Match the words with the same vowel sound.

1	chest	☐	calf
2	arm	1	head
3	heel	☐	knee
4	face	☐	thigh
5	eye	☐	throat
6	nose	☐	waist

Now listen and check. Say the words aloud.

VOCABULARY

1 Underline the parts of the body.

ankle | back | bald | big | brown | ear
elbow | fat | finger | fingernail | foot | forehead
hair | leg | lip | long | mouth | neck | shoulder
slim | stomach | thumb | toe | tooth | wrist

2 Complete the chart with the adjectives in activity 1 and parts of the body they can describe. You can use the parts of the body in *Sounds.*

adjectives	parts of the body
bald	hair
big	stomach, foot, mouth
brown	hair, eye
long	hair, finger, leg, body, nose
slim	body, wrist, neck,
fat	stomach, finger,

3 Which parts of the body can you use to complete this sentence?

I've got two ...

elbows, feet, shoulders, wrists, eyes, ears, hands, ankles, thumbs, lips, legs

_____

_____

GRAMMAR

1 Match the questions and answers. Put the correct numbers in the boxes.

1 Have you been to university?
2 Have you ever been in a play?
3 Have you ever been to a rugby match?
4 Have you been to the dentist this month?
5 Have you ever been married?
6 Have you been to a rock concert?

☐ Yes, I had an appointment last week.
☐ No, I haven't. I don't like acting.
☐ Yes, I have. But I'm divorced now.
1 No, I haven't. I left school when I was sixteen.
☐ No, I haven't. I prefer classical music.
☐ Yes, and I've played too.

2 Complete the chart of irregular verbs.

infinitive	past simple	past participle
become	became	become
break	_____	_____
drink	_____	_____
drive	_____	_____
eat	_____	_____
pay	_____	_____
run	_____	_____
say	_____	_____
speak	_____	_____
teach	_____	_____
wear	_____	_____

3 Complete the questions with past participles from the chart in activity 2.

1 Have you ever _eaten_ frogs' legs?
2 Have you ever _____ an arm or a leg?
3 Have you ever _____ mint tea?
4 Have you ever _____ to a politician?
5 Have you ever _____ five kilometres?
6 Have you ever _____ a hire car?

4 Answer the questions in activity 3. Give more information about your experiences.

1 _____
_____

2 _____
_____

3 _____
_____

4 _____
_____

5 _____
_____

6 _____
_____

5 Your answers are *Yes, I have* and *No, I haven't*. Write questions. Don't use the verbs in activity 3.

Yes, I have.

_____

_____

_____

No, I haven't.

_____

_____

_____

6 Complete these questions with *Did*, *Have* or *Has*.

1 *Did* _____ you study languages at university?

2 _____ there been an election recently?

3 _____ you ever lived in the United States?

4 _____ you leave home before you went to college?

5 _____ you like primary school?

6 _____ Jack Nicholson made a new film?

LISTENING

1 📼 Listen to four people talking about their health. They are each answering one of the questions below. Write the number of the speaker by the question they answer.

Have you ever had a dreadful cough? ☐

Have you ever had an accident while playing a sport? ☐

Have you ever been ill on holiday? ☐

Have you ever given up something because of your health? ☐

Have you ever been ill for three or four weeks? ☐

Have you ever cut yourself badly? ☐

2 What are their answers to the questions?

Speaker 1 _____

Speaker 2 _____

Speaker 3 _____

Speaker 4 _____

📼 Listen again and check.

3 Here are the answers to some more questions about the speakers' health. Write the questions.

1 _____

Yes, I did. I went to hospital in an ambulance – there was one there because it was a race. I had to have eight stitches.

2 _____

Yes, I have. I think I've taken them every time I've been to Latin America. But I've never had malaria so I'm not complaining.

3 _____

Hardly ever. I live on my own, you see – so I don't get colds or flu, or anything more serious, from other people.

4 _____

Yes, I think I probably am. I go running two or three times a week, and I play tennis and squash. Yes, I'm quite fit.

📼 Listen and check.

22 | *What's new with you?*

GRAMMAR

1 Complete the sentences with *'ve* or *'s*.

1 He *'s* travelled a lot for his job.
2 We ____ drunk all the tea.
3 She ____ got divorced.
4 I ____ had a party.
5 You ____ changed your pen.
6 They ____ written a couple of times.

2 Complete the sentences with *'ve* or *'s* where possible.

1 I *'ve* bought some new clothes.
2 They ____ came home late.
3 She ____ found a new job.
4 He ____ rang his parents from Tokyo.
5 We ____ voted twice.
6 You ____ finished this exercise.

3 Complete the chart.

infinitive	past participle
feel	felt
_____	given
_____	gone
_____	heard
_____	known
_____	left
_____	met
_____	read
_____	seen
_____	spoken
_____	taken
_____	thought

4 Look at the past participle forms in the chart in activity 3 again. Which past participle forms are the same as their past simple forms? Write a list.

_____ _____ _____

_____ _____ _____

5 Complete the sentences with the simple past or present perfect form of the verbs in brackets.

1 I don't know when he _*died*_. (die)
2 My sister _____ a baby in July so now she's got three children. (have)
3 Oh, no! I _____ my bag. (lose)
4 'How long _____ you _____ Bert?' (know)
   'Since 1970.'
5 '_____ you ever _____ Italy?' (visit)
   'Yes, I _____ there last year.' (go)
6 It's much colder than it _____ last year. (be)
7 'When _____ you _____ your job?' (change)
   'In October.'
8 My colleague travels a lot. She _____ in some of the best hotels in the world. (stay)

6 Read this paragraph about Andy. Use the verbs in brackets to write sentences about things that have and haven't changed in Andy's life.

> *I'm still living at home but my girlfriend and I are going to get married next month so we've bought a flat of our own. It's really nice. I'm still in the same job too – but I took some exams last month so perhaps I'll be able to get a better job with opportunities for travel abroad. Talking of travel, I visited Turkey recently. I went there for a two-week holiday. It was great!*

1 (buy) _____
2 (change) _____
3 (get) _____
4 (move) _____
5 (take) _____
6 (visit) _____

LISTENING

1 🔊 Listen to four people talking about things they have done recently. Write the number of the speaker by the things they talk about.

☐ lived somewhere

☐ moved somewhere

☐ passed something

☐ played something

☐ started something

☐ stopped something

2 Write down what each speaker has done in more detail.

Speaker 1 _____

_____

_____

Speaker 2 _____

_____

_____

Speaker 3 _____

_____

_____

Speaker 4 _____

_____

_____

🔊 Listen again and check.

SOUNDS

Read this true sentence.

Her brother has bought a house.

🔊 Listen and correct the statements below with the true sentence. Change the stressed word each time.

1 His brother has bought a house.

2 Her sister has bought a house.

3 Her brother has rented a house.

4 Her brother has bought a car.

VOCABULARY AND WRITING

1 Think about recent news in your country. Choose and underline six of these topics.

education   election   employment   finance
freedom   government   housing   inflation
law and order   police   rights
standard of living   tourism   war

2 Write the six topics you chose in activity 1 in a list below and add words you associate with each one.

1 _____

_____

2 _____

_____

3 _____

_____

4 _____

_____

5 _____

_____

6 _____

_____

3 What's happened recently? Write sentences about the news topics you chose in activity 1. Use the present perfect tense.

1 _____

_____

2 _____

_____

3 _____

_____

4 _____

_____

5 _____

_____

6 _____

_____

4 Choose someone you know well. Write a paragraph about this person. Describe some things that have and haven't changed for this person in the last year.

It's a holiday

SOUNDS

Write these words in three groups. Underline the stressed syllables in the words with more than one syllable.

celebrate flag holiday king parade
picnic president queen soldiers

one syllable	two syllables	three syllables
flag	_____	_____
_____	_____	_____
_____	_____	_____

▭ Now listen and check. Say the words aloud.

VOCABULARY

1 Complete these sentences with words from one of the groups in *Sounds*.

1 On St Patrick's Day there is a
_____ through New York.

2 There have been British _____ in Northern Ireland for more than twenty years.

3 We sometimes go for a _____ in summer.

2 Choose one of the other groups in *Sounds*. Write sentences about your country with the three words.

1 _____
   _____
2 _____
   _____
3 _____
   _____

GRAMMAR

1 What time, day, month or year is it now? Rewrite the expressions with *since*.

1 for five minutes
   since twenty-five past four
2 for a quarter of an hour _____
3 for three hours _____
4 for twenty-four hours _____
5 for four days _____
6 for two weeks _____
7 for five months _____
8 for six years _____

2 Rewrite these expressions with *for*. Which is the shortest length of time? Which is the longest? Put the lengths of time in order by numbering the boxes – 1 = shortest, 8 = longest.

since Saturday *for three days* _____ ▢

since January 1st _____ ▢

since the summer holidays _____ ▢

since 8.00am _____ ▢

since May _____ ▢

since last Wednesday _____ ▢

since 1990 _____ ▢

since Christmas _____ ▢

3 Write eight true sentences about your friends and family. Use *for* and *since* and eight of the expressions in activities 1 and 2.

1 *My brother has been up since 8.00am.*
2 _____
3 _____
4 _____
5 _____
6 _____
7 _____
8 _____

READING

1 You are going to read about St Valentine's Day. What is the connection between St Valentine's Day and the following?

- anonymous cards
- fertility festival
- saint of lovers
- third century
- 14th February

2 Read the text about St Valentine's Day and check your answers:

St Valentine's Day

St Valentine's Day is celebrated on 14th February. Today, St Valentine's Day provides a chance to let someone know you are interested in them, or for a secret admirer to declare his or her love to you.

Our present-day Valentine customs come from a mixture of pagan and Christian tradition and ancient folk beliefs. St Valentine was probably a priest killed on 14th February in the time of the Roman Emperor Claudius in the third century AD. He became known as the saint of lovers because he performed forbidden marriage ceremonies for Roman soliders.

But the Romans had celebrated 14th February for centuries as Lupercalia. This was a fertility festival when young people tried to find a partner.

In one eighteenth-century Valentine custom, a man chose a girl's name from a hat. The name he chose gave the name of his future wife and he wore the name pinned to his shirt sleeve for the next few days.

The first known Valentine card was sent in 1415 when a French duke, who was locked in the Tower of London, sent the first rhyming love-letter to his wife. Valentine cards first went on sale in the early nineteenth century.

3 Read the sentences below. Complete each paragraph in the text with one of the sentences.

a They weren't supposed to marry in case it made them less able to fight.

b This is probably where we get the expression 'to wear your heart on your sleeve'.

c It is, in fact, an excuse for any sort of romantic celebration or exchange.

d The Christian church of the fourth century used the tradition of St Valentine to take the pagan feast of Lupercalia and give it Christian meaning.

e Today, the custom is to send anonymous cards.

LISTENING

1 🔲 Listen to people talking about St Valentine's Day. Put a tick (✓) if the speakers have received Valentine cards and a cross (✗) if they haven't.

Speaker 1 ❑

Speaker 2 ❑

Speaker 3 ❑

2 Write down any other things the speakers have done on St Valentine's Day.

Speaker 1 _____

_____

Speaker 2 _____

_____

Speaker 3 _____

_____

🔲 Listen again and check.

SPEAKING

1 You are going to talk about what you have and haven't done on St Valentine's Day. Think about what you're going to say.

2 Record yourself on a blank cassette.

GRAMMAR

1 Find the definitions and put the correct numbers in the boxes.

1 A cashpoint is a machine which...
2 A traffic warden is someone who...
3 A newsagent's is a shop which...
4 An L-driver is a person who...
5 A bottle bank is a container where...
6 A double yellow line is a mark on the road which...

6 ✓ means 'no parking'.

5 ✓ you can put used glass items for recycling.

1 gives money to bank customers.

2 ✓ gives parking tickets when drivers break the rules.

3 ✓ sells newspapers and magazines.

4 ✓ is learning to drive.

2 Look at the picture. Put a tick (✓) next to the numbers of the things in activity 1 that you can see in the picture.

1 ☒ 2 ☑ 3 ☒ 4 ☑ 5 ☒ 6 ☑

3 Replace *that* with *who* or *which* in the following sentences.

1 A folk museum is a museum that/ *which* exhibits historic items of everyday use.
2 A crammer is a private school that/ *which* ✓ prepares students for examinations.
3 A lady-in-waiting is a woman that/ *who* ✓ attends members of the royal family. ✓
4 A shamrock is a plant that/ *which* ✓ is the national emblem of Ireland.
5 Shepherd's pie is a hot, savoury dish that/ *which* ✓ contains meat, potato and carrots.
6 A newsagent is someone that/ *who* sells newspapers.

4 Complete the sentences with *who, which* and *where*.

1 I know someone *who* can speak five languages.
2 Did you get the message *which* ✓ I left for you?
3 I like words *which* ✓ sound nice.
4 The chemist's is the shop *where* ✓ you can buy medicine.
5 Pilots are people *who* ✓ fly planes.
6 Let's go *where* ✓ we went last week.

5 Complete these sentences so that they are true for you. Use *who, which* or *where*.

1 I know someone _____
   _____
2 I've got a friend _____
   _____
3 I've never been to a country _____
   _____
4 I know a place _____
   _____
5 I don't like cars _____
   _____
6 I've never had a holiday _____
   _____

SOUNDS

📼 Listen to these sentences. Are the speakers British or American? Write B or A.

1 How fast can you drive?
2 August's the hottest month.
3 Who's that man with red hair?
4 I can't stand picnics and barbecues.
5 My brother works in New York.
6 How much do rooms cost?

LISTENING

1 Match the American words on the left with their British English equivalent on the right.

American	British
closet	sweets/chocolate
first floor	cupboard
to call collect	aubergine
eggplant	rubbish
elevator	ground floor
to stand in line	film
movie	single ticket
garbage	lift
one-way ticket	to queue
round-trip ticket	to make a reverse charge call
truck	return ticket
candy	lorry

2 📼 Listen to an American and an English person talking about differences in vocabulary. Did you match the words correctly in activity 1?

3 How did the speakers define each word?

📼 Listen again and check.

READING

1 Read the newspaper article and choose the best title.

1 How to speak American
2 The Latin States of America
3 Spanish-speaking America
4 The languages of America

With the exception of English, Spanish is now the most common language spoken in the United States according to a report from the United States Census Bureau.

The report says that the number of Spanish speakers increased by about 50 per cent between 1980 and 1990. This brings the total number to over 17 million.

In some neighbourhoods of New York City, Miami (with a 60 per cent Spanish-speaking population) and many Texan cities, you don't hear much English at all. In some cities you often see signs and advertisements in Spanish.

The Census Bureau says that one in seven US residents speaks a language other than English at home. This figure has risen from 23.1 million a decade ago to 31.8 million.

Three out of four people who said that they spoke a foreign language at home also said that they spoke English very well. Some communities have their own TV and radio stations, newspapers and schools. In public schools in many states, children have their lessons in a variety of languages, including, for example in California, Armenian, Cantonese, Japanese, Korean, Russian and Spanish, as well as English.

But the fastest-growing language in the US is Mon-Khmer, the language of Cambodia; the number of residents who speak Mon-Khmer at home has grown by more than 676 per cent over ten years.

2 Do people in your country speak as many languages as are spoken in the United States? Which are the most important languages spoken in your country?

_____

3 Does any information in the article surprise you?

_____
_____
_____

VOCABULARY AND LISTENING

1 Look at the four photos. Which of these words describe the things?

big cardboard cloth cotton curved flat
glass hard heavy leather light long metal
oval oblong paper plastic nylon round
soft small square thick thin wood wool

A *flat* _____
B _____
C _____
D _____

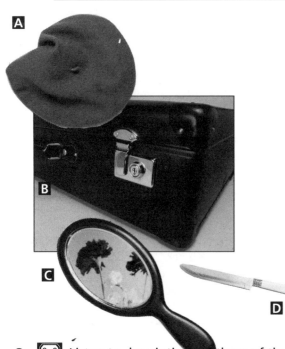

A

B

C

D

2 🔊 Listen to descriptions of three of the objects. Match the descriptions and the objects and put the correct letters in the boxes.

1 ☐ 2 ☐ 3 ☐

3 🔊 Listen again. Do the descriptions include the words you chose in activity 1?

4 Complete this description of the fourth object.

It's made of _____ and _____ .

It's _____ , _____ and

_____ .

SOUNDS

1 Put the words in the right order and make questions.

1 look what it like does
  What does it look like? _____

2 it soft is

  _____

3 made it what's of

  _____

4 it is oblong

  _____

5 very is it big

  _____

6 how it does weigh much

  _____

2 🔊 Listen and notice the words the speaker links. Say the sentences aloud.

GRAMMAR

1 Think of an object you own. Answer the questions in *Sounds* activity 1.

1 _____
2 _____
3 _____
4 _____
5 _____
6 _____

2 Complete these descriptions with the names of the objects in the pictures in *Vocabulary and listening* activity 1.

1 You use a __*knife*__ to cut food.

2 A _____ is for looking at yourself.

3 You use a _____ to keep your head warm.

4 A _____ is for carrying things.

3 Rewrite the sentences in activity 2.

1 A _knife is for cutting things._

2 To _____

3 A _____

4 To _____

4 Describe these things and their purpose.

1 A comb _____

_____

2 A notepad _____

_____

3 A phonecard _____

_____

4 A towel _____

_____

5 A tent _____

_____

6 A wallet _____

_____

5 Look at your Student's Book and complete these sentences.

1 To refer to the grammar notes for the present simple, you turn to page ____ .

2 To revise lessons 36-40, you do the activities on pages ____ and ____ .

3 To do the communication activity in Lesson 14, Student *B* turns to page ____ .

4 To find out about Guy Fawkes Night, you read the text on page ____ .

5 You look at page ____ to check the past tense of *see*.

6 Write similar sentences about your Practice Book.

1 _____

_____

2 _____

_____

3 _____

_____

4 _____

_____

READING AND LISTENING

1 Look at the three words below. For each word, there are two false definitions and one true one. Read the definitions and choose the true one.

1 anorak

a A large, oblong box often made of metal which you use to carry your belongings, clothes, books and things, especially when you're going to live somewhere else for a long time.

b A warm coat which you wear in winter. It's often made of nylon and is waterproof. It's a fairly short coat – it comes to the top of your legs.

c A machine you use to cut the grass. It runs on electricity or petrol, and has a box at the back to collect the grass.

2 eiderdown

a A large, soft thing made of cotton and filled with feathers which you use on your bed to sleep under and to keep warm at night.

b A pair of trousers used for work, made of denim or some other material and which also cover the chest. They have braces over the shoulder to hold them up.

c A small piece of cloth which you use to wash your face.

3 spanner

a A tool you use when you're working on a car. It's a long, metal thing with two round bits at either end where the nuts go.

b A thing you give to a baby to keep it quiet. It's made of plastic or rubber and the baby puts it in its mouth.

c Stuff you use to clean ovens. It's a thick white foam which you spray on to the oven while it's still warm and then wipe off with a wet cloth or sponge.

2 Listen to the true definitions. Did you guess correctly in activity 1?

3 The false definitions in activity 1 are true definitions for the following words. Put the number and letter of the definition by the word. There is one extra word.

dummy _____ dungarees _____

face cloth _____ lawn mower _____

oven cleaner _____ pushchair _____

trunk _____

4 Listen to the definitions. Did you guess correctly?

5 Write two false definitions and one true definition for the extra word in activity 3.

SOUNDS

Match the words with the same vowel sound.

1	lane	☐	head
2	door	☐	out
3	down	☐	throw
4	get	1	train
5	go	☐	walk
6	use	☐	you

🔊 Now listen and check. Say the words aloud.

GRAMMAR

1 Match the rules with the road signs. Put the correct numbers in the boxes.

☐ You have to turn left ahead.

☐ You mustn't turn right.

☐ You mustn't turn round.

1 You have to give priority to vehicles from the opposite direction.

☐ You have to stop in 100 yards.

☐ You mustn't go down this road.

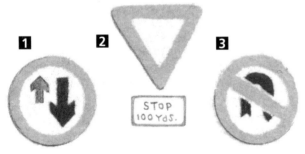

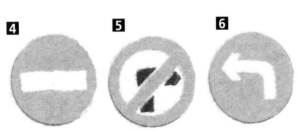

2 Complete the rules for each road sign with *have to* or *mustn't*.

1 You *have to* slow down.

2 You _____ stop.

3 You _____ keep left.

4 You _____ allow other vehicles to go first.

5 You _____ pass other vehicles.

6 You _____ go more than 30 miles per hour.

3 Complete these sentences with *must* or *mustn't*.

1 The doctor says my brother *must* give up smoking.

2 You _____ work harder or you won't pass your exams.

3 You _____ throw anything from the train window.

4 I haven't got a film in my camera. I _____ buy one.

5 I've got a dentist's appointment and I _____ be late.

6 Nice to meet you! We _____ have lunch one day. Goodbye.

4 Add *to* where necessary in these sentences.

1 Do I have _to_ change trains?

2 Would you like _____ sit next to a window?

3 We're going _____ book our tickets tomorrow.

4 You mustn't _____ smoke on internal flights in the United States.

5 Try not _____ cross the road between parked vehicles.

6 Do not _____ drive on the right in Britain.

VOCABULARY AND READING

1 You are going to read some advice for road users on foot. Which of these words do you expect to find in the text?

bicycle carriage clear cross customs drive
duty-free fast first class footpath guard lane
look pavement policeman ride road
second class speed ticket inspector traffic walk

2 Read the advice for road users on foot. Underline the words from activity 1 in the text.

3 Which pieces of advice are the people in the pictures not following? Write the numbers in the boxes.

Picture 1 ☐

Picture 2 ☐

Picture 3 ☐

Picture 4 ☐

4 Tell the people in the pictures what they must and mustn't do.

1 You must _____
_____.

2 You mustn't _____
_____.

3 You must _____
_____.

4 You must _____
_____.

WRITING

Give some advice for getting on and off a bus.

Road users on foot

a Where there is a pavement or suitable footpath, use it. Do not walk on the edge of the pavement with your back to the traffic.

b Where there is no suitable footpath, keep as close as possible to the side of the road and face oncoming traffic. Walk behind not next to one another. Take care at bends.

c If you have children with you, walk between them and the traffic. Do not let them run into the road.

d You can be more easily seen in the dark or in poor light if you wear or carry something white, or light-coloured, or reflective. This is important on country roads without footpaths.

e Before you cross a road, stop at the edge of the pavement and look both ways. When the road is clear, walk straight across but keep looking out and listening for traffic. Cross the road as quickly as you can, but do not run.

VOCABULARY

1 Complete these extracts from *The Skylight* with verbs from the list below. Use the correct form of the verbs.

asleep axe climb door farm grass
hammer hill hurry knock ladder lock
lower narrow roof shout shutter skylight
sleep smash square stones suitcase
thumb tools toys wave whisper

1 The driver tried the door. It was *locked* _____.
2 She _____ down the ladder.
3 She climbed up the ladder again and

_____, 'Johnny, can you hear me?'
4 When she saw the lights of the car she

_____ her arms to stop it.
5 She _____, 'Oh, thank God.'

2 The verbs you used in activity 1 can also all be nouns. Write sentences using the same words as nouns. (You can use a dictionary.)

1 _____
2 _____
3 _____
4 _____
5 _____

3 Find fifteen verbs in the puzzle. They go in two directions: → and ↓. Use each letter once only.

D	R	I	N	K	U	S	E
T	S	T	A	N	D	D	S
O	R	P	L	A	Y	R	L
U	U	D	R	A	W	I	E
C	N	R	I	D	E	V	E
H	C	L	I	M	B	E	P
R	E	A	D	M	A	K	E
S	P	E	A	K	S	E	E

GRAMMAR

1 Complete the sentences with *can, can't, could* or *couldn't* and nine of the verbs in the puzzle.

1 I _____ _____ on my hands.
2 I _____ _____ English ten years ago.
3 I _____ _____ a truck.
4 I _____ _____ pictures when I was a child.
5 I _____ _____ coffee without sugar.
6 I _____ _____ trees when I was small.
7 I _____ _____ very well without glasses.
8 I _____ _____ tennis ten years ago.
9 I _____ _____ my toes.

2 Use the other six verbs to write questions.

1 Can you _____
2 Could you _____
3 Can you _____
4 Could you _____
5 Can you _____
6 Could you _____

3 Complete the sentences with *can* or *can't*.

1 I go to the swimming pool every week, but I *can't* swim very well.
2 My spoken French isn't very good, but I _____ read it quite well.
3 He _____ see very easily at football matches because he isn't very tall.
4 I'm not bad at running and I _____ cycle too.
5 Please speak up! I _____ hear you.
6 I've always been right-handed – I _____ do anything with my left hand.

4 These are your answers. Write questions.

1 Yes, I can.

_____

2 No, I can't.

_____

3 Yes, I could.

_____

4 No, I couldn't.

_____

SOUNDS

Say these words aloud.

a /kæn/   b /kən/   c /kɑːnt/

🔊 Listen to these sentences and decide if you hear a, b, or c.

1 Who can give me the answer? | b |

2 Sorry, I can't. ☐

3 Can you hear me? ☐

4 Yes, I can. ☐

5 I can see you too. ☐

6 I can't tell him. ☐

Now say the sentences aloud.

READING AND WRITING

1 You're going to read an article about Dr Ron Parise who spent nine days in the space shuttle *Columbia*. Write down five or six daily routine activities. Which ones do you think you can do in space? Which ones are more difficult to do?

_____

_____

_____

Now read *Just an ordinary day?* and find out if Ron Parise mentions the daily routine activities you thought of.

2 Do you know someone with an unusual job? Think of someone you know or have heard or read about. Describe this person's day.

In 1990, Dr Ron Parise was one of a team of astronauts who went on a nine-day mission that took them round the Earth 143 times. Dr Parise describes life on board the space shuttle Columbia.

JUST AN ORDINARY DAY?

'Living and working in zero gravity is OK. It's quite nice floating around, but you have to fix your feet into position while you're working so that you don't float off.

But it wasn't all work. There was lots of time for serious things like standing on the ceiling and holding each other up with one finger. One of our favourite games was throwing and catching sweets and blowing water at each other.

There are some problems with living in space. For example, it's pretty difficult to eat chips or spaghetti. Food is freeze-dried and kept in sealed boxes. It's not bad, but you know you're not going into space to eat gourmet meals. It's also a bit like hospital because you have to choose all your meals beforehand.

Cleaning your teeth is a problem. You can't brush your teeth with your mouth open because it's a very messy job. If you spill anything in the shuttle, you have to catch it in a cloth otherwise it just keeps floating around.

Astronauts have twelve hours of hard work and then twelve hours off. You have to rest in your bed even if you don't sleep. Each sleeping compartment is very narrow and closed like a long cupboard. It's also vertical – or horizontal depending on which way you decide is up!'

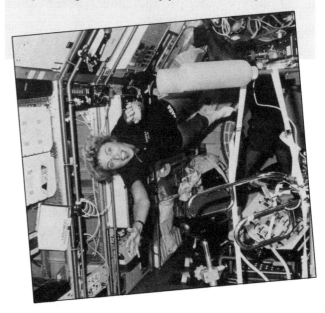

28 | *Breaking the rules?*

READING AND WRITING

1 Read these grammar rules. Are they true or false? Write T or F in the boxes.

Grammar quiz

1 You can use the present simple to talk about what is happening at the moment. ☐

2 You can put an adverb of frequency between the noun/pronoun and the verb. ☐

3 You can't use the future simple to make predictions. ☐

4 You can use the present perfect simple to talk about something that happened in the past and which affects the present. ☐

5 You can say *When have you stayed in hospital?* ☐

6 You can use *going to* to talk about future intentions or plans which are fairly certain. ☐

7 You can't put an *-s* ending on a modal verb. ☐

8 You can't use an *-ing* form verb after *like*. ☐

9 You can form all plurals by adding *-s*. ☐

10 You can use *tomorrow* and *next...* to talk about the past. ☐

11 You can use *who, what,* and *which* to ask about the subject and the object of the sentence. ☐

12 You can say *colder, fitter, expensiver, intelligenter.* ☐

13 You can't use the contracted form in spoken and informal written English. ☐

14 You can't use the definite article with place names. ☐

15 You can use *at* to talk about times of the day. ☐

16 You can't say *I'm student of English.* ☐

17 You can't put *do* with the verb *be*. ☐

18 You can begin a letter with *Dear...* ☐

19 You can't put a comma after *but*. ☐

20 You can say *she's a very handsome little girl.* ☐

21 You can't say *I've lived here since twenty years.* ☐

22 You can use *that* to replace *who* and *which*. ☐

23 You can use *to* + infinitive to describe the purpose of something. ☐

24 You can't use *get* to mean *become*. ☐

25 You can't use *must* in the past. ☐

2 Rewrite the false grammar rules so that they are true. Use *can* and *can't*.

GRAMMAR

1 Write full answers to these questions about rules in your country.

1 When can you start school?

_____

2 When can you get married?

_____

3 When can you drive a car?

_____

4 When can you have your own passport?

_____

5 When can you vote?

_____

6 When can you leave school?

_____

2 Look at the signs above and say what you can or can't do.

1 _____
2 _____
3 _____
4 _____
5 _____
6 _____

THESE SEATS ARE MEANT FOR ELDERLY AND HANDICAPPED PERSONS
Please give up your seat to someone who is less able to stand than you

3 Complete these sentences to make true statements about your country.

1 You can cross the street _____

_____

2 You can't cross the street _____

_____

3 You can drive at 90 kilometres an hour

_____

4 You can't drive at 90 kilometres an hour

_____

5 You can go to university _____

_____

6 You can't go to university _____

_____

7 You can have a drink in a bar _____

_____

8 You can't have a drink in a bar _____

_____

4 Write three sentences saying what you can do when you're sixteen in your country, and three sentences saying what you can't do.

_____

_____

_____

_____

_____

_____

SOUNDS

Listen to the way speaker B uses a strong stress and intonation to sound insistent.

1 A Can I go now?

B You know you can't.

2 A I can't do this.

B You can.

3 A Can I walk on the grass?

B No, you can't.

4 A I'm sure we can walk there.

B We can't! It's too far.

Say the sentences aloud. Try to sound insistent like speaker B.

VOCABULARY AND LISTENING

1 Where could you hear these sentences? Choose places from the list.

1 'You can cook your own meals.'

2 'You can take five books out at a time.'

3 'You can have an alcoholic drink but you have to pay for it.'

4 'You can't take photos of the exhibits.'

aeroplane   bar   beach   bus   canal   church
escalator   hospital   hotel   level crossing   library
lift   mosque   mountain   museum   office   park
pedestrian crossing   petrol station   prison   shop
stadium   traffic lights   train   youth hostel   zoo

Listen and check your answers.

2 Listen again. What else can you and can't you do in these places?

1 _____

_____

2 _____

_____

3 _____

_____

4 _____

_____

57

READING AND WRITING

1 You are going to read a passage about American sport. Before you read it, try to answer these questions. Write full answers, but leave out anything you don't know or cannot guess.

1 What are the most popular sports in North America?

The most popular sports in North America are
_____

2 Who are the spectators of these sports?
_____
_____

3 When is the baseball season?
_____
_____

4 When is the football season?
_____
_____

5 Why are college teams important?
_____
_____

2 Read *Spectator sports* and complete or correct your answers to the questions in activity 1.

3 Write down the words from the passage that you can use to talk about sports.
_____
_____
_____
_____

4 Write a paragraph about sport in your country. Use *Spectator sports* and the words you wrote down in activity 3 to help you.

Spectator sports

Many North Americans love sport. This does not necessarily mean that they get any exercise. What they do is watch national teams on television. They watch baseball, football, basketball, hockey, golf and tennis – which means that for most of the weekend they sit in front of the television. If they are sociable, their friends watch with them.

Baseball is the great American sport and has fans in every income and ethnic group. Nearly every major city has a team, and as each team plays 162 games a year, following baseball can take up a lot of time. The season is spring and summer and finishes in the fall World Series, when the two leading teams play each other. The first to win four games wins the Series. Although called the World Series, Canada is the only other country to participate.

Football, played in fall and winter, is also very popular. It is not at all like the game called soccer. There isn't much kicking in American football, which has more to do with running with the ball and knocking people down. Serious injuries are common in football. Each team only plays 16 games during the regular season, and these take place on Sundays and Monday evenings. Half the country watches the final game on Superbowl Sunday, in January.

When there are no professional games on television, Americans watch college sports. Many colleges feel that if they want their former students to give money to their old school, they need a winning football team. Curiously, this seems to be true.

Some adults continue to play baseball after high school and college on amateur teams. Although these games are supposed to be purely for fun, Americans care a lot about winning!

SOUNDS

Put these sentences in the correct order and make a dialogue.

a   Is it very far?

b   Thanks very much.

c   Excuse me. Can you tell me how to get to the bus station?

d   No, about half a mile. I'm going that way myself.

e   Yes, of course. Jump in.

f   Go straight on. Then take the second left after the traffic lights. You can't miss it.

g   Could I have a lift?

c						

Listen and check. Do you think the speakers sound polite and friendly?

Say the conversation aloud. Try to sound friendly and polite.

GRAMMAR

1   Put a tick (✓) if B gives permission or a cross (✗) if he/she refuses permission in these situations.

1   A   Can I watch TV?

    B   Yes, go ahead.   ❑

2   A   Could I use your phone?

    B   By all means.   ❑

3   A   May I get off here?

    B   I'm afraid you can't.   ❑

4   A   Is it all right if we sit here?

    B   I'm sorry, those seats are taken.   ❑

5   A   Would you mind if I had a drink?

    B   No, go ahead.   ❑

2   Write your own responses to these requests.

1   I wonder if you could lend me some money?

    _____

2   Could you tell me the time, please?

    _____

3   Would you mind writing your address?

    _____

4   Can you write more clearly, please?

    _____

5   Please could you tell me where the list of irregular verbs is?

    _____

6   Could you help me with my homework, please?

    _____

3   Complete the offers.

1   Can someone answer the phone?
    I'll   *answer it.* _____

2   I need some stamps.
    Shall I _____

3   How about a cup of tea?
    I'll _____

4   I haven't got time to do this.
    Shall I _____

5   Who's going to book the tickets?
    I'll _____

6   I can't find my keys.
    Shall I _____

4   Ask for permission or ask someone to do something. Write in the speech bubbles.

5 Write down six useful sentences for asking people to do things in your English lessons.

1 _____
2 _____
3 _____
4 _____
5 _____
6 _____

6 Write down six useful sentences for asking permission in your English lessons.

1 _____
2 _____
3 _____
4 _____
5 _____
6 _____

LISTENING

1 Listen to four conversations between a teacher and her students. Write P if the students ask for permission and D if they ask the teacher to do things.

1 ☐ 2 ☐ 3 ☐ 4 ☐

2 Do the students ask any of the things you wrote in *Grammar* activities 5 and 6? What else do they ask?

1 _____
2 _____
3 _____
4 _____

3 What does the teacher say in reply?

1 _____
2 _____
3 _____
4 _____

Listen again and check your answers.

VOCABULARY AND WRITING

1 Write two more situations for the *Doing things the right way* questionnaire in your Student's Book. Try to use some of the words in the list.

borrow   date   diary   formal
get on with someone   helping   hold on
informal   lend   manners   neighbour
phone number   polite   rude   share   turn down

_____
_____
_____
_____
_____
_____
_____
_____
_____
_____
_____

2 Read your situations again. Decide what you would do in those situations.

31 | *My strangest dream*

READING

1 Read what Ben Noakes says about his interest in ghosts from the introduction to his book *I saw a ghost*. Which of these sentences explains what happened? Tick (✓) the box.

1 Ben saw a ghost. ☐

2 Ben heard a ghost. ☐

3 His father saw a ghost. ☐ ✓

4 His father heard a ghost. ☑

2 Which of these times is most important in the story? What happened at these times?

three o'clock

Father and mother heard a ghost

seven o'clock

he go back to his party

ten o'clock

he wake up

3 Have you ever seen or heard a ghost? Do you know anyone who has? Do you believe in ghosts?

My interest in ghosts and the supernatural began about seven years ago when I was fourteen. A friend of mine invited me to a Christmas party on 13th December. I remember little about the party itself, but it went on until well into the early hours – the early light was breaking through as I got into a taxi for my journey home. I tipped the driver far too much, climbed the stairs to my bedroom and slept.

My father awoke me at about ten o'clock.

'Three o'clock in the morning is no time for a party,' he said. 'Why did you bring your friends back here?'

What was he talking about? A party here?

'I went out to a party, Dad,' I protested. 'I didn't get back until about seven this morning.' Father looked puzzled.

'Come on, now – your mother and I both heard your voice. There was no doubt about it. And we heard other voices too.'

My father noted the incident in his diary, and didn't say another word about it – until 14th December the following year.

I was staying for a few days with a school-friend in the country so there was no doubt in my parents' minds that I was out of London. At about three o'clock in the morning, my parents woke up when they heard my voice – they are certain of this – in the garden under their bedroom window. They could hear me calling for the family dog, 'Tramp! Tramp!' Downstairs in the kitchen, Tramp was barking. My father climbed out of bed and looked out of the window. The noise stopped. All he could hear was the sound of dead leaves flying in the wind, and the distant bark of the dog.

It was only when he wrote his diary that my father realised that this incident had taken place at three o'clock in the morning of 14th December – exactly a year to the hour since what we now call 'The Ghosts' Party'.

GRAMMAR

1 Look at the pictures of Kate opposite. What was she doing at the times in the pictures?

1 *She was having breakfast at eight o'clock.*
2 _____
3 _____
4 _____
5 _____
6 _____

2 What were you doing yesterday at the times shown in the pictures in activity 1? Write sentences.

1 _____
2 _____
3 _____
4 _____
5 _____
6 _____

3 Match the two parts of the sentence. Then put the verbs in either the past simple or past continuous to make six sentences with *when*.

1 I (cross) the road   I (hurt) my ankle.
2 I (watch) TV   it (start) to snow.
3 I (record) a film   the car (hit) me.
4 I (run) in a race   there (be) a power cut.
5 I (get) some money   the video (break) down.
6 I (ski) without a hat   the gunman (come) into the bank.

1 *I was crossing the road when the car hit me.*
2 _____
3 _____
4 _____
5 _____
6 _____

SOUNDS

Listen to these sentences. Notice the stress and intonation in sentences with two parts.

1 I was having lunch when the phone rang.
2 He was driving quite fast when the accident happened.
3 It was snowing when we left home.
4 We weren't watching when the child fell.
5 I was living in London when I met him.
6 She was working abroad when her father died.

Now say the sentences aloud.

VOCABULARY

1 Make ten two-word nouns with these words.

bag bite chicken closing insect instrument jet lag limit military mobile musical officer phone police roast service sleeping speed time

closing time _____
_____ _____
_____ _____
_____ _____

2 How many two-word nouns can you make using these words?

1 art _____
2 boarding _____
3 car _____
4 cheese _____
5 guide _____
6 railway _____
7 single _____
8 swimming _____
9 traffic _____
10 traveller's _____

SOUNDS

1 Say these words aloud. Which of the words have the sound /ə/? Underline the sound.

disappear   discover   evil   favourite   grounds
prison   shiver   strange   warn

🎧 Now listen and check.

2 Put the words in activity 1 into four groups according to the number of syllables they have.

one syllable	two syllables	three syllables
grounds	_____	_____
_____	_____	_____
_____	_____	_____

🎧 Now listen and check. Say the words aloud.

VOCABULARY

1 Divide the words in *Sounds* activity 1 into three groups.

nouns	*grounds*
adjectives	_____
verbs	_____

2 Complete the sentences with the words in *Sounds* activity 1. There are three extra words.

1 When did Alexander Fleming *discover* penicillin?

2 You can visit the palace and walk in its _____ .

3 I quite like rock 'n' roll, but my _____ type of music is jazz.

4 The thieves went to _____ for five years.

5 Why do we _____ when we're cold?

6 My brother phoned to _____ me about the train strike.

GRAMMAR

1 Put the words in order and write sentences. Remember to separate the clauses with a comma (,).

1 we it church while to the standing started rain were outside

While we were standing outside the church, it started to rain.

2 notes the while taking the was ran policewoman away thieves colour

While the policewoman was taking notes, the thieves ran away ✓

3 the mother school working while children stayed at was their

While their mother was working, the children stayed at school

4 my was arrived having I while cousins lunch

While I was having lunch, my cousins arrived ✓

5 shower the he a telephone was rang while having

While he was having a shower, the telephone rang ✓

6 were my the while his cousin arm climbing mountain broke we

While we were climbing the mountain, my cousin broke his arm ✓

2 Match these sentences with the situations in activity 1. Put the number of the situation in the correct box.

[2] She ran after them and caught one of them. The other one got away.

[6] A helicopter came and took him to hospital.

[5] It stopped ringing before he could get to it.

[1] We put up our umbrellas and ran to the car.

[3] She collected them at about five o'clock when she left the office.

[4] I asked them if they would like something to eat.

3 Rewrite these sentences using *while*.

1 They were travelling in France when they heard the news.

   While they were travelling in France, they heard the news.

2 When she came into my office, I was speaking on the phone.

   ~~She came into my office~~ while ~~I was speaking on the phone~~ ✓

3 When he got sunstroke, he was playing tennis.

   While ~~he was playing tennis, he got sunstroke.~~ ✓

4 He was playing against his old team when he scored his first goal.

   ~~He scored his first goal~~ while ~~he was playing against his old team~~ ✓

4 Write sentences saying what happened next in the situations in activity 3.

1 ~~They were waiting in a traffic jam when they took a coming out.~~

2 ~~When I finished my conversation, she was sitting and was weating patiently.~~

3 ~~When the doctor came, he was sitting in his bed~~

4 ~~When the game finished, he was having some compliments~~

5 Underline the action that happened first.

1 I made a cup of tea when <u>I got home</u>. ✓

2 She became Queen when <u>her father died.</u>

3 When <u>I sat down</u>, I suddenly felt very tired.

4 I looked at the dinosaurs first when <u>I went to the museum.</u> ✓

5 She fell asleep when <u>the lights went out.</u> ✓

6 After <u>she arrived,</u> she felt tired.

6 Complete the passage with the past simple or the past continuous tense of the verbs.

Robert Bateman, internationally-acclaimed artist and naturalist, (1) <u>was working</u> (work) on a painting of a lioness and her cub and (2) <u>wanted</u> ✓ (want) to portray them in as natural an environment as possible. He (3) <u>remembered</u> (remember) an old apple tree not far from his rural Canadian home. It was just like some trees in Africa he (4) <u>knew</u> ✓ (know). He (5) <u>packed</u> (pack) his painting equipment in the car and (6) <u>prepared</u> (prepare) to go. But first he (7) <u>drew</u> ✓ (draw) the lioness and cub on the canvas in pencil.

While he (8) <u>was working</u> ✓ (work), a farmer and his son (9) <u>stopped</u> ✓ (stop) to watch him. When he (10) <u>saw</u> ✓ (see) their surprise at what he (11) <u>was painting</u> ✓ (paint), Bateman (12) <u>said</u> ✓ (say), 'Too bad! They were here half an hour ago. You've just missed them.'

SPEAKING

1 You are going to tell a story in English. Choose a funny story you have heard. Think about what you're going to say.

2 Record yourself on a blank cassette.

VOCABULARY

1 Write the opposites of these adjectives.

1 dirty   *clean*
2 dry   _____
3 fewer   _____
4 hilly   _____
5 poor   _____
6 quiet   _____

GRAMMAR

1 Complete the chart.

nouns	adjectives
1 heat	*hot*
2 hill	_____
3 industry	_____
4 peace	_____
5 rain	_____

2 Complete each sentence with one of the words with the same number in activity 1.

1 It isn't _____ enough in our house in winter.
2 There are too many _____ for cyclists.
3 There isn't enough _____ in the town.
4 It isn't _____ enough for the baby to sleep.
5 There's too much _____ to go out for a walk.

3 Rewrite the sentences with the other words in the chart in activity 1.

1 *There isn't enough heat in our house in winter.*
2 _____
3 _____
4 _____
5 _____

4 Complete these complaints about a town in Britain with *enough, too much/many*. Begin the sentences with *There is/isn't* or *There are/aren't*.

1 *There aren't enough* car parks.
2 _____ litter on the streets.
3 _____ tourists in summer.
4 _____ traffic.
5 _____ nightlife.
6 _____ people out of work.

5 Read your sentences in activity 4. Which sentences are true for your village, town or part of the city? Put a tick (✓) next to the numbers of the sentences which are true.

1 ☐   2 ☐   3 ☐   4 ☐   5 ☐   6 ☐

6 Rewrite these sentences with *fewer* and *less*.

1 There are more cinemas than theatres in Britain.
   _____

2 There is more industry in Wales than in Northern Ireland.
   _____

3 There is more unemployment in the north of England than in the south.
   _____

4 There are more motorways today than twenty years ago.
   _____

5 There is more rain in winter than in summer.
   _____

SOUNDS

Underline the stressed words.

It's quite industrial in the south of Wales and more people live there. The north is poorer and it's more rural. It's quite peaceful in the north. There are some mountains in the north-west and it's hilly in the centre. There's quite a lot of farmland in Wales. It's a beautiful country.

Listen and check. Read the passage aloud.

1 *T* Support more rural bus routes, extend bus lanes and encourage the design of buses which are suitable for disabled people.

2 Oppose large-scale mineral extractions.

3 Introduce planning guidelines to stop developments which encourage car use (eg out-of-town shopping).

4 Recycle 60 per cent of domestic waste by the year 2000 instead of burying it and encourage the door-to-door collection of rubbish.

5 Provide more footpaths and road crossings for pedestrians.

6 Pay council workers' travel expenses at the same rate for walking, cycling or driving a car.

7 Introduce energy conservation measures in its own buildings and include solar heating in the design of new buildings.

8 Encourage the use of recycled building materials.

READING

1 Read about some of the things the Green Party in Britain would like to do. Decide if the things are connected with transport or with conservation. Write T or C in the boxes.

2 Which suggestions do you think are good suggestions? Put a tick (✓) next to the good suggestions.
1 ☐ 2 ☐ 3 ☐ 4 ☐
5 ☐ 6 ☐ 7 ☐ 8 ☐

LISTENING

1 Listen to four people talking about improving our environment. Decide if the things they suggest are connected with transport or with conservation. Write T or C.
Speaker 1 ☐
Speaker 2 ☐
Speaker 3 ☐
Speaker 4 ☐

2 Make notes about the suggestions.
Speaker 1 _____
_____
Speaker 2 _____
_____
Speaker 3 _____
_____
Speaker 4 _____
_____

3 Listen again and check. Which suggestions do you think are the best?

READING

1 Write words you associate with Carnival.

_____

_____

_____

2 Read the passage. How many of your words are in the passage?

3 Write captions for the photos.

1 _____
_____

2 _____
_____

3 _____
_____

Notting Hill Carnival

The carnivals in Rio de Janeiro, New Orleans and the Caribbean islands are famous all over the world for their brilliant parades of dancers in costumes and exciting music. Carnivals started many years ago as Christian festivals which began the period of Lent. They are held in January or February – winter months in the countries where Christianity first developed.

For countries like Brazil and the islands of the Caribbean the weather is warm in January and February. When people from the Caribbean came to live in Britain in the 1950s they found that Carnival occurred in the coldest part of the British winter. Although they celebrated Carnival with parties in their homes and churches, they missed the fun of their traditional open air summer celebrations.

Many of the people from the Caribbean live in an area of London called Notting Hill. In 1961 they decided to bring back their festival and held the first Notting Hill Carnival at the end of August. Although it lost its Christian origins and dates, it was a great success, and it is now held every year. The Notting Hill Carnival is now one of London's most exciting festivals. As at other carnivals throughout the world, there are several parades of decorated vehicles and dancers in costumes. In the side streets there are parties until late in the night, with plenty of food and music.

4 Make a list of other festivals which are celebrated at different times of the year in your country.

Spring

_____

Summer

_____

Autumn

_____

Winter

_____

SOUNDS

Look at this true sentence.

The shops are closed on Monday mornings.

Listen and correct the statements below with the true sentence. Change the stressed word each time.

1 The schools are closed on Monday mornings.

2 The shops are open on Monday mornings.

3 The shops are closed on Tuesday mornings.

4 The shops are closed on Monday afternoons.

GRAMMAR

1 Underline the six verbs with irregular past participles. Write the participles.

allow believe build call celebrate eat give hold know name write use

built _construie_   eaten   given

held   known   writen

2 Complete this description with the passive form of five of the verbs in activity 1.

In Britain the first day of Lent (1) _is known_ as Ash Wednesday. The day before Ash Wednesday (2) _is called_ Shrove Tuesday. Shrove Tuesday is traditionally the last day of celebration before Lent; it (3) _is celebrated_ 40 days before Easter. Pancakes (4) _are eaten_ on Shrove Tuesday so its popular name is Pancake Tuesday. In some towns pancake races (5) _are held_ .

3 Complete the sentences about Easter in Britain with the active or passive forms of these verbs.

close   decorate   eat   exchange   give   sell

1 Most churches _are decorated_ with flowers for Easter Sunday.

2 Most shops _are closed_ on Good Friday and Easter Monday.

3 People ____eat____ eggs at Easter to mark the birth of new life and the beginning of spring.

4 Chocolate Easter eggs _are sold_ in the shops.

5 Some people _exchang ed_ greeting cards and chocolate eggs.

6 Children _are given_ one or two weeks' holiday.

4 Rewrite these sentences in the passive.

1 In Britain they call Easter buns 'hot cross buns'.
In Britain Easter buns are called 'hot cross buns'.

2 People eat hot cross buns on Good Friday.
Hot cross buns are eaten on Good Friday by people

3 They make the buns with dried fruit.
The buns are made with dried fruit.

4 They mark the buns with a cross before they bake them.
The buns are marked with a cross before they are baked

5 They toast the buns before they eat them.
The buns are toasted before they are eaten

6 They serve the buns with butter.
The buns are served with butter

WRITING

Write a description of something which is eaten in your country on a special occasion.

READING

Are you a good tourist? Do the quiz and find out.

Are you a good tourist?

Check your knowledge of international etiquette with this holiday travel test.

1 CYPRUS: On a village tour, a local offers you a *glyko*. What should you do?
a Ride it – it's a type of bicycle.
b Wear it – it's a traditional hat.
c Eat it – it's a sweetened fruit dessert.

2 MOROCCO: You'd like to pay a visit to some mosques, but are you allowed inside?
a Yes, if you take your shoes off.
b No, unless you're a Muslim.
c Yes, except during Ramadan.

3 JAPAN: You're staying in a Japanese inn and decide to relax in a traditional, shared bath. What shouldn't you do in the bath?
a Wash yourself.
b Talk to other bathers.
c Stay too long.

4 NEPAL: While trekking, you want to photograph local villagers. Is this OK?
a Yes – the Nepalese love it.
b No – it's against their religion.
c Yes – but ask permission first.

5 GERMANY: After you walk across a street, you're stopped and fined by the police. Why?
a You ignored the pedestrian crossing.
b The pedestrian light was red.
c You didn't look both ways before you crossed.

6 SINGAPORE: You offer a piece of chewing-gum to your tour guide. Why does she look at you in such a strange way?
a Chewing-gum is banned in Singapore.
b Tour guides aren't allowed to accept gifts.
c Chewing-gum is only given to animals.

7 INDIA: An Indian friend invites you to a traditional meal. Which is the correct way to eat?
a With your left hand only.
b With your right hand only.
c With both hands.

8 AUSTRALIA: In a bar, you observe some angry Australians. What's the most likely reason?
a Australia has lost at cricket.
b Their beer is warm.
c The beer glasses are warm.

9 RUSSIA: You are introduced to some Russians. When you greet them, what should you avoid?
a Shaking hands as you enter their home.
b Extending your left hand.
c Shaking hands with your gloves on.

10 CHINA: If you're invited into someone's house, which of these actions may cause offence to your hosts?
a Blowing your nose.
b Refusing an offer of food.
c Not removing your shoes before entering.

Key:

Score one point for each question correctly answered.

1 c Glyko is a traditional dish, given in welcome by villagers. It is impolite to refuse it or offer money for it.
2 b Non-Muslims are banned from Moroccan mosques.
3 a You should wash yourself at a tap before relaxing in the bath.
4 c As in many other cultures, the Nepalese are sensitive about being photographed.
5 a and b You must use crossings correctly even if there is no traffic.
6 a People are fined in Singapore for chewing gum.
7 b Indians traditionally eat with their right hand; the left is considered unclean.
8 a, b and c It's a trick question!
9 a It's considered very unlucky.
10 b It's considered impolite to refuse food, although people usually refuse before they accept.

How did you do?

8–10 Congratulations! You've completed your round-the-world tour with the minimum of embarrassment.
5–7 Not bad, but you need to take a bit more care.
2–4 Have you ever wondered why locals look at you as if you're crazy?
0–1 Try a holiday at home this year.

Adapted from Holiday Which?, published by Consumers' Association and reproduced with their permission.

GRAMMAR

1 Match the two parts of the sentences and put the correct numbers in the boxes.

1 I don't drink a lot of milk,

2 Although there aren't many good restaurants near our home,

3 We've got cups and saucers,

4 Pasta in Italy is always a starter,

5 I sometimes miss lunch,

6 I usually have cereal for breakfast,

☐ but I always have a good dinner.

☐ but we usually drink from mugs.

☐ we eat out about once a week.

☐ although I quite like toast occasionally.

[1] although I use it in puddings.

☐ although we usually have it as a main course.

2 Rewrite these sentences. Use the words in brackets.

1 I like chocolate a lot. But I try to have only one piece a day. (although)
Although I like chocolate a lot, I try to have only one piece a day.

2 I have a sweet tooth. However, I try not to have too much sugar in my tea. (but)

_____

_____

3 I like salad with meat or fish but my husband prefers cooked vegetables. (although)

_____

_____

4 Although we have a light breakfast during the week, we have a large brunch on Sundays. (however)

_____

_____

5 Although we quite like pasta, we don't have it very often. (but)

_____

_____

SOUNDS

Match the words with the same vowel sound.

1 chips ☐ saucer

2 fork ☐ slice

3 knife ☐ spoon

4 plate ☐ steak

5 soup [1] dish

🔊 Now listen and check. Say the words aloud.

VOCABULARY

1 Look at the picture. How many of these things can you see? Tick (✓) the boxes.

bowl ❑ cup ❑ dish ❑
fork ❑ glass ❑ knife ❑
napkin ❑ plate ❑ saucer ❑
spoon ❑ teaspoon ❑ table cloth ❑

2 What are the people eating? Write two lists.

the woman	the man
fish	

36 | *Lovely weather*

GRAMMAR

1 Complete the sentences with *so* or *because*. Remember to use a comma (,) before *so*.

1 I've forgotten my umbrella ,_so_____ I'm going to get wet.

2 Ecuador has the same weather all year round _____ it's on the equator.

3 We can't go skiing today _____ it's snowing really hard.

4 I can see lightning _____ don't stand under that tree.

5 We're having our holiday in May _____ it's the best time to go to the States.

6 It'll be cold at night _____ let's take our sleeping bags.

2 Write what day of the week, month or year it will be:

1 tomorrow _____

2 the day after tomorrow _____

3 next month _____

4 in three months _____

5 next year _____

6 in two years' time _____

3 Write sentences with *will/won't, may/may not* or *might/might not*. Use the six expressions of time in activity 2.

1 _____
    _____

2 _____
    _____

3 _____
    _____

4 _____
    _____

5 _____
    _____

6 _____
    _____

VOCABULARY

1 Look at the list of words and the three photos. Write words for each photo. You can use some words more than once but it isn't necessary to use all the words.

changeable dry flood fog freezing frost humid hurricane ice lightning mild mist rain snow storm sun thunder wet wind

1 nouns: _*flood*_____
    adjectives: _____

2 nouns: _____
    adjectives: _____

3 nouns: _____
    adjectives: _____

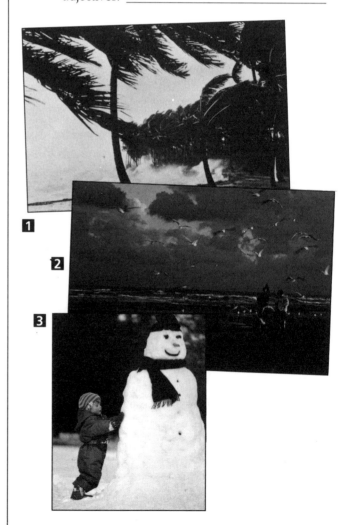

READING AND WRITING

1 Read the weather forecast. Circle the words you can use to talk about the weather.

FORECAST FOR WEDNESDAY 10TH JUNE: After a (misty) start, most places will brighten up with some sunshine. However, much of the east coast will stay cloudy. Rainy periods are possible in all parts, but are most likely in southern parts, Wales and Northern Ireland. Some showers will be heavy with thunder. It will be warm with temperatures 22°C, that's 72°F inland, and 16°C - 61°F on the east coast.

2 Match the weather forecast with the correct map of Britain.

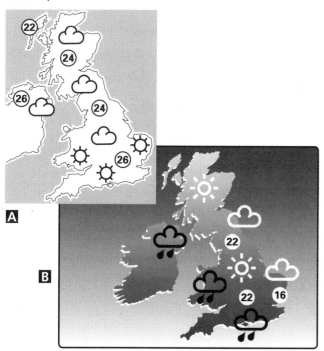

A

B

3 It is the morning of Wednesday 10th June. Are these sentences true or false? Write T or F in the boxes.

1 Most parts of Britain will have mist all day. ☐ *F*

2 There may be thunder in Wales. ☐

3 It will probably be dry in the south of England. ☐

4 It might be cloudy all day in the west. ☐

5 It will be warmer on the east coast. ☐

6 The sun may shine all day in Northern Ireland. ☐

4 Look at the other map. It shows the weather for Thursday 11th June. Write the forecast.

_____
_____
_____
_____
_____
_____
_____

SOUNDS

Look at the weather forecast in *Reading and writing* activity 1. Underline the stressed words.

🔊 Now listen and check. Read the forecast aloud.

LISTENING

1 Imagine you are on holiday in Britain. It is the first Friday in August. What words do think you might hear in a weather forecast in August?

_____
_____

2 🔊 Listen to the weather forecast for the weekend. How many of your words did you hear?

3 🔊 You want to spend one day in museums and one day at a safari park. Listen to the forecast again and decide which thing to do on Saturday and which to do on Sunday. Give reasons.

_____
_____
_____

WRITING

Look in a newspaper. Find a weather forecast for your country for the next two days. Write predictions about the weather in English.

73

37 | *Help!*

READING

1 Read the passage quickly and decide what kind of emergency the speaker is describing. Choose from the following.

an emergency at home   a road accident
a bomb alert   a mugging   a flood

2 Decide where these sentences can go in the story.

1 'If we break down the door, we'll cause a lot of damage,' the fire chief said.

2 'If the water doesn't stop, it'll ruin my furniture and carpets,' I thought.

3 'If we come, we'll charge you £48,' the man said.

4 'But if we don't stop the water, it will cause problems with the electricity.'

5 'If I ring him, at least he'll have a key,' I thought.

GRAMMAR

1 Complete the sentences with the correct forms of the verbs in brackets.

1 If I ~~did~~ *do* ~~will be~~ (do) well in my exams, I *will be* (be) pleased.

2 If I *fail* (fail) my exams, I *won't be able* (not be able) to go to university.

3 If I *don't have* (not have) breakfast at home, I *will buy* (buy) something to eat.

4 If we *don't get* (not get) the first train, we *will wait* (wait) for the next one.

5 If I *drink* (drink) this, *will* I *feel* (feel) better?

6 If there *is* (be) a train strike tomorrow, I *won't go* (not go) to work.

I was tired after work, and I was looking forward to sitting down and spending a relaxing evening at home. But when I walked in the door, I noticed water running through the ceiling.

(a) _____ I ran upstairs and knocked on the neighbours' door, but there was no answer. I decided to ring the landlord, in case he knew where they were.

(b) _____ But he was out too. By now the water was everywhere. So I called the water authorities, but they were too expensive and I didn't have enough money.

(c) _____ In the end I called the fire brigade.

'In theory we should charge you for this,' said the fire chief. (d) _____ Let's call it an emergency.'

So the youngest firefighter walked round to the back of the house, climbed up a ladder and balancing dangerously on the roof, tried to open the skylight. But it was stuck.

(e) _____ I think we'll just see if the landlord is back before we do that.' By chance he was home now, and he said it was OK to break in. 'But could you break a smaller window, please? There's one at the front.' The fire chief agreed, and soon they were inside, and found the cause of the flood, a burst hot water tank. I was amazed by how careful and thoughtful they were, even though it was an emergency.

74

2 Write sentences. What will happen if you... ?

1 wake up late tomorrow morning

_____

_____

2 have a stomach upset tomorrow evening

_____

_____

3 forget to do your homework

_____

_____

4 have to work at the weekend

_____

_____

5 don't remember a friend's birthday

_____

_____

6 miss your favourite TV programme

_____

_____

SOUNDS

1 🔊 Listen to these phrases. Notice how you don't always hear the '*ll* sound.

1 I leave   I'll leave
2 we listen   we'll listen
3 they learn   they'll learn
4 I lose   I'll lose
5 we like   we'll like
6 they look   they'll look

2 Now look at the phrases in context and underline the correct verb form.

1 If *I leave*/*I'll leave* soon, I'll get there on time.
2 If *we listen*/*we'll listen*, we'll find out what has happened.
3 If they work harder, *they learn*/*they'll learn* more.
4 If *I lose*/*I'll lose* my job, I'll look for another one.
5 If *we like*/*we'll like* the car, we'll buy it.
6 If we paint the walls, *they look*/*they'll look* nice.

🔊 Listen and check. Say the sentences aloud.

VOCABULARY

1 Complete the sentences with some of these words or expressions. You may need to modify nouns and verbs.

accident   ambulance   bomb   break   burglar
burn   button   catch fire   consulate   dangerous
drown   electricity   explode   fire   flood   injured
gas   ground floor   gun   kill   mugger   plug
plug in   press   rescue   shock   steal   switch off
switch on   unplug   victim   wallet   witness

1 Moving an _*injured*_ person can be very _____ .

2 A _____ with a _____ took the old man's _____ while he was getting money out of the cashpoint machine.

3 Some people _____ their _____ and _____ when they go on holiday but I only _____ my television.

2 Write sentences with some of the other words in activity 1.

_____

_____

_____

_____

_____

LISTENING

1 Imagine there's a fire in your home and you can't get out. Which of these things do you think you should do? Tick (✓) the boxes.

1 close doors   ❑
2 go to a window   ❑
3 go down to floor level ❑
4 open the window   ❑
5 call for help   ❑
6 wait for the fire brigade ❑

2 🔊 Listen to a fire officer talking about what you should do. Did you agree with her?

38 | *My perfect weekend*

SOUNDS

Look at these words. Which ones contain the sounds /k/ , /s/ or /ʃ/ ? Put them in three columns.

cassette cat champagne comb companion computer dog insurance jewellery juice novel pen religion rice salt seafood shoes shopping soap telephone water

words with /k/	words with /s/	words with /ʃ/
cassette	*cassette*	_____
_____	_____	_____
_____	_____	_____
_____	_____	_____
_____	_____	
		_____

Listen and check. Say the words aloud.

VOCABULARY

1 Complete the sentences with six of the words in *Sounds*.

1 A good *novel* is a necessity for me because I always like to read before I go to sleep.

2 I don't think a _____ is a necessity or a luxury. It would only chase my cats!

3 I think _____ is an essential part of a Sunday breakfast. I make my own with fresh oranges.

4 I think a pencil is more essential than a _____ – you can't rub ink out.

5 A _____ is a necessity for my job. I've got a mobile so I can speak to colleagues and clients while I'm in my car.

6 I think _____ is a necessity. We need it to wash with.

2 Write sentences about six things that are or are not necessities and luxuries for you. Use six of the other things in *Sounds*.

1 _____
2 _____
3 _____
4 _____
5 _____
6 _____

3 Look at the photos of these two holiday destinations. Write words which you associate with each type of holiday.

_____
_____

_____
_____

GRAMMAR

1 You are discussing an imaginary winter holiday with a friend. Put the words in order to ask your friend questions.

1 would go place to you which
Which place would you go to?

2 would that one choose you why
Why would you choose that one?

3 you when go would
When would you go?

4 with who would take you you
Who would you take with you?

5 spend your how holiday would you
How would you spend your holiday?

2 Choose one of the holiday photos and answer the questions you wrote in activity 1.

1 *I would go to Chamonix*
2 *I would choose that one because I like ski*
3 *I would go for the next Wint holiday.*
4 *I would take my family and my friend.*
5 *I would go skiing and walking*

LISTENING

1 Listen to four people talking about the things below. Write the numbers of the speakers by the things they talk about.

☐ perfect car ☐ perfect day
☐ perfect holiday ☐ perfect home
☐ perfect meal ☐ perfect office

2 What do the people say? Use *he* or *she*.

1 _____
2 _____
3 _____
4 _____

Listen and check.

READING AND WRITING

1 Here are some answers to questions about *My perfect weekend,* which you read in Student's Book Lesson 38. Write the questions.

1 *What would you read?*
I'd like the chance to re-read some of the novels by Anthony Burgess. I think I'd start with *Earthly Powers*.

2 *Which CD would you take?*
Some Schubert, Mahler's *Song of the Earth,* and Elgar's cello concerto. They will all make me think of my family and friends.

3 *What would you regret from yours home? Would you watch T.V.?*
Probably nothing. I only really enjoy television at home. When I'm abroad, I can't manage to understand what's going on.

4 *Would you buy some postcards?*
Yes, I'd send one to my mother. She loves receiving postcards from foreign countries!

5 *What would you do?*
I'd play a little tennis, if I found someone who wasn't too good. And I expect I'd go for my morning run, and walk a great deal.

6 *What would you miss when you were away?*
My children and my home. I would miss them while I was away, so even after a perfect weekend, I'd look forward to getting back.

2 Write your own answers to the questions in activity 1.

1 _____
2 _____
3 _____
4 _____
5 _____
6 _____

VOCABULARY

1 Write words which you associate with the words in the list below. You can use your dictionary to help you.

cafe   dentist   exchange   expect   favour
happen   gentleman   note   pleased
put on   run   silk   stand still   suspicious
take out   taxi   umbrella   unpleasant

_____

_____

_____

_____

_____

GRAMMAR

1 Complete these sentences with the past tense or *would* + infinitive. Use these verbs:

be   borrow   go   have   ~~help~~   ~~leave~~   ~~meet~~
offer   take   tell   try   want

1 If I _went_ to New York, I _would take_ ✓ a boat to the Statue of Liberty.

2 If I _(were)_ was tired, I _would have_ an early night.

3 If I _met_ ✓ a famous person, I _would tell_ ✓ all my friends about it.

4 If I _left_ ✓ my wallet at home, I _would borrow_ ✓ some money from my friends.

5 I _would help_ my friends with their homework if they _wanted_ me to.

6 I _would try_ frogs' legs if someone _offered_ me them.

2 Read the sentences in activity 1 again. Would you do the same? Write yes or no in the boxes.

1 [no]   2 [yes]   3 [no]

4 [yes]   5 [yes]   6 [no]

3 In what circumstances would you do these things? Complete the sentences.

1 I would hire a car if... _I left home_ ✓ _____

2 I would learn Japanese if... _I went to Japan_ ✓

3 I would move house if... _I had a lot of children_ ✓

4 I would use a dictionary if... _I didn't know a word_

5 I would carry my passport if... _I went to an other continent_ ✓

6 I would phone the doctor if... _I were sick_ ✓

4 Which of these things are likely to happen to you? Which are unlikely? Write L or U in the boxes.

1 watch TV this evening   [L]

2 move house   [U]

3 get a job in the USA   [U]

4 break your leg   [U]

5 don't do all your homework   [L]

6 have a birthday party   [U]

5 Write sentences about the six things in activity 4. Use *If* to introduce each thing and then write about the result.

1 _If I watch TV this evening, I wouldn't finish my homework_

2 _If I move house, I wouldn't be happy._

3 _If I break my leg, I would be in hospital._

4 _If I get a job in the USA, I wouldn't take it._

5 _If I don't do all my homework, I would be punished._

6 _If I have a birthday party, I wouldn't eat so much._

SOUNDS

1 🔊 Listen to these phrases. Notice how you don't always hear the *'d* sound.

1 I travel   I'd travel
2 we do   we'd do
3 they take   they'd take
4 I dance   I'd dance
5 you drive   you'd drive
6 they tell   they'd tell

2 Now look at the phrases in context and underline the correct verb form.

1 *I travel/I'd travel* abroad if I had the money.
2 If *we do/we'd do* our homework, we'll be able to watch TV.
3 If I were you, *I take/I'd take* an umbrella.
4 If someone at a party asked me, *I dance/I'd dance* with her.
5 We'll have an accident if *you drive/you'd drive* so fast.
6 If they knew the answer, *they tell/they'd tell* you.

🔊 Listen and check. Say the sentences aloud.

READING

1 You're going to read part of a story called *Mean Streets*. First, answer the questions.

What would you do if ...

1 you were in a strange city and you had nowhere to stay?

_____

_____

2 it was late and you only had traveller's cheques?

_____

_____

3 someone offered to help you change some money and find a hotel?

_____

_____

Now read the story.

Mean Streets

Calvin was a tall man, and wore a leather jacket. He told me he played semi-professional football during the day. We met one evening at the Greyhound bus station in San Francisco. I was looking in the telephone directory for a hotel nearby. I wanted to move on to Stanford but the last bus had left, and I needed somewhere to stay the night. I saw him talking to another man out in the street, and there seemed to be some kind of problem. Then Calvin came over to me and asked me to change $50.

'I only have traveller's cheques,' I replied. I explained my situation.

He said, 'I'll take you to a hotel, they can cash a cheque for you, you can give me some change, and we'll all be happy.'

'You need me out there,' said Calvin. 'Those streets, it's like New York.' He picked up the bigger of my bags and we walked into the streets. Nothing reminded me of England.

'A lot of people would charge you for protection, but you and me are friends.' He sounded sincere. 'You're safe with me. I've got a gun.'

The Hotel Oxford was full. The Hotel Bristol didn't take traveller's cheques. The New London Hotel turned us away. Next door was a McDonald's. Calvin greeted the staff and ordered two drinks. I paid with a $100 cheque and got $97.21 in change. He then said, 'Give me $50 and I'll go back to the Greyhound station and give him the change.' I hesitated.

He looked at me. 'Man, don't worry,' he said, putting a hand on my shoulder. 'If I wanted your money, I would take it right now. You and me are friends. Listen. You ring me tomorrow and I'll drive you to Stanford.'

I gave him a $50 note and watched him walk out into the streets. Even before he reached the door, I knew I wouldn't see him again. I was 20 metres from the station and $50 poorer. I felt like a baby.

2 Would you do the same as the writer? If not, at what point in the story would you do things differently?

3 In the same circumstances, what would you do next? Write the end of the story.

_____
_____
_____
_____
_____
_____

79

GRAMMAR AND READING

1 Put these sentences in the right order to make a story.

a After he had helped her climb out of the expensive vehicle, he asked her, 'What do you think?'

b 'But hopefully when your business improves, you'll be able to afford something with four doors.'

c 'It's not bad, I suppose,' she replied.

d He decided to take her for a ride.

e A successful young businessman had just bought a new Porsche and wanted to show it off to his grandmother.

| e | ☐ | ☐ | ☐ | ☐ |

2 Write the full forms of these contracted verbs. Is *'d* the contracted form of *would* or *had*?

1 he'd arrest   *would*

2 she'd crashed _____

3 I'd had   _____

4 we'd mend   _____

5 they'd packed _____

6 she'd queue   _____

7 I'd spent   _____

8 we'd want   _____

3 Underline the verbs in the past perfect tense.

1 <u>I'd cut</u> my finger so I went to the doctor.

2 It'd cost much more if you flew there.

3 She'd hit the policeman when he arrested her.

4 He'd put everything away unless you didn't want him to.

5 If I were you, I'd read a magazine.

6 He'd set off before he looked at the map.

4 Underline the action which happened first.

1 We got there on time but <u>our friends had already left</u>.

2 Her brother had become ill before she got there.

3 I bought these shoes after I had tried them on.

4 The train left when the guard blew his whistle.

5 The party had started when they arrived.

6 When I saw the dog I ran as fast as I could.

5 Complete the sentences with the past perfect tense of suitable verbs.

1 I didn't do my homework because I'd *left* my books at home.

2 The children didn't want to go to the cinema because they'd already _____ the film.

3 Katie wasn't at home last week because she'd _____ to her uncle's.

4 My mother never knew her father because he'd _____ before she was born.

5 I was quite excited when the plane took off because I'd never _____ before.

6 My grandfather was always a bit afraid of animals because he had always _____ in cities.

6 Read the passage and underline the correct form of the verbs.

Robert Falcon Scott, the most famous of British polar explorers, was born in Devonport, Britain in 1868. Because he had failed to reach the South Pole in 1908, Scott (1)*decided/had decided* to make a second attempt. When Scott (2)*chose/had chosen* 40 men to take part in the expedition, his ship, the *Terra Nova*, set off from Plymouth. After they had picked up supplies in New Zealand, the *Terra Nova* reached Antarctica on 1st November 1911 and the men (3)*made/had made* their base. Scott then (4)*asked/had asked* four men to accompany

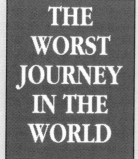

THE WORST JOURNEY IN THE WORLD

him on the final part of his journey. One of the four men, Edward Wilson, (5)*was/had been* with him on his first expedition. The five men finally set off for the Pole after bad weather and illness had delayed them for two months. On 16th January 1912, when they got to the Pole, they (6)*found/had found* a Norwegian flag there. Scott's great rival, Roald Amundsen, (7)*arrived/had arrived* a month before them. Tragically, Scott and his companions (8)*didn't take /hadn't taken* enough food for their return journey to the *Terra Nova* and all five men died.

VOCABULARY

Write sentences with the verb forms you didn't use in *Grammar and reading* activity 6. Use another verb (either in the simple past or past perfect) and at least one of the words or expressions in the list in each sentence.

accident   arrest   break down   capital   cliff   cost   crash   exhaust
fall off   guard   journey   mend   musician   nightmare   pack   pipe
police   queue   rusty   set off   spend

1 *He had decided to leave, when the police arrived.*
2 _____
3 _____
4 _____
5 _____
6 _____
7 _____
8 _____

SOUNDS

Notice the stress and intonation in sentences with two parts.

1 After she'd had time off work, she felt better.
2 When he'd had his lunch, he went out for a walk.
3 Because she'd worked so hard, she did really well in her exams.
4 After he'd switched on the computer, he started work.
5 Because he'd run ten kilometres, he was very tired.
6 When he'd sold his old car, he bought another.

Now say the sentences aloud.

LISTENING

1 ▶ Listen to Mary talking about an incident in France. Why did the policeman stop the car? Choose from these reasons.

1 They were driving on the wrong side of the road.
2 They hadn't stopped at the road junction.
3 They hadn't paid the motorway toll.
4 They had gone the wrong way.

2 What had Mary and her friends decided to do before they set off on their journey? Why did Mary have an argument with one of her friends? Has she spoken to him since?

▶ Listen again and check.

WRITING

Have you ever been very lucky or very unlucky on a journey you have made? Describe the journey.

Tapescripts

Lesson 2 Listening and writing, activity 1

Um... I get up at about half past seven. I have three children and they have to get off to school. They're five, seven and eight years old so I take them to school after breakfast. We leave home at about half past eight. The school's not far – we walk there. When I get back home, I always have a cup of coffee and have a look at the newspaper. Then I start work. I'm a writer – I write for a magazine and I work at home. I have an office with my computer and I sit there most of the day. I usually stop for lunch at half past twelve and then I continue until about half past three when the children come home from school. My husband gets home from work at, er... about a quarter to six so we usually eat at six o'clock. We all have dinner together. I often read or watch television after the children are in bed. It's a nice, quiet time. I go to bed quite early myself – usually at about half past ten. And that's my day.

Lesson 3 Listening and writing, activity 1

MAN	Do you live in a house or a flat?
WOMAN	I live in a flat – a ground-floor flat.
MAN	What's the main room in your home?
WOMAN	Well, the biggest room's the living room, so that's the main room. It's not an enormous flat, you see. There's a living room, a kitchen, a bathroom and then there are the bedrooms.
MAN	How many bedrooms are there?
WOMAN	Two bedrooms. One's my bedroom and, um... the other one's for visitors. I have quite a lot of people to stay. I like cooking, so, um... I'm quite happy when people come to see me. I've got a wonderful kitchen with a deep freeze, an oven and lots of other electrical equipment.
MAN	Have you got a dishwasher?
WOMAN	Oh, yes! I like cooking but I, um... don't like washing dishes! It's not a very big kitchen, so I have to put the plates and things in the dishwasher after every meal.
MAN	Where do you eat your meals?
WOMAN	In the living room. There isn't a table in the kitchen. As I say, the living room's the main room.
MAN	Do you have a television?
WOMAN	I have two – one in the living room and the other in my bedroom!

Lesson 6 Listening, activity 1

Yes, I read the book Paul Theroux wrote about his journey around the coast of Britain. It's called *The Kingdom by the Sea*. He's actually American but I think he lived in London. Yes, his wife was British. Anyway, he started his journey on May 1st, that's, er... May 1st 1983 – that's when he left London and his home. He took a train to Margate, that was the first place he visited. And after that he took another train to Deal, that's where he spent his first night. He stayed in a bed and breakfast for his first night. I seem to remember he was the only guest. Then he continued along the south coast of England, through Dover, Hastings, Eastbourne and Brighton. He met lots of people and in Hastings he spent quite a lot of time with some artists. A man called Mr Bratby painted his portrait in Hastings. And then he went on down to the south-west, Devon and Cornwall, and then up into Wales. When he was in Wales, he visited the famous British travel writer, Jan Morris. Remember, Theroux's a famous American travel writer. Yes, he visited Jan Morris in Criccieth where she lives. And after that he went on to Liverpool, Carlisle, Glasgow and on to Stranraer. He took a boat to Northern Ireland from Stranraer. I can't remember how long he spent there, in fact, I can't remember how long the journey took him. Anyway, he returned to Scotland and continued on his way, all around the coast of Scotland. He met lots of people on his journey, but not many people knew he was a famous writer. He did meet one woman who knew his books. She worked in a bookshop in St Andrews, and she told him that the Queen was in St Andrews. He nearly saw the Queen but, in fact, he didn't see her. Anyway, the book ends when he gets to Southend. I suppose he then went back to London.

Lesson 7 Listening and writing, activity 1

Speaker 1 It was in January. That's January this year. It's not a bad time for a skiing holiday. I try to go most years.

Speaker 2 I went with two friends. They're friends from university. The three of us often go on holiday together.

Speaker 3 We went to Turkey – to the south-west, not to Istanbul. I'd like to go there some time, though.

Speaker 4 The thing I liked best was going down into the Grand Canyon. It was fantastic! We stayed in a hostel in the bottom of the canyon for two nights.

Lesson 9 Listening, activity 2

WOMAN	This is a photo of the new man in my life. What do you think?
MAN	He looks quite nice. What's his name?
WOMAN	James – James Smith. Quite an ordinary name really.
MAN	And this is him at work?
WOMAN	Yes, he gave me the photo. I didn't take it. He's got his own company. He sells doors and windows and things like that. You don't need any new windows, do you?
MAN	No, not really. He's got his own company... How old is he?
WOMAN	He's only 32 – or that's what he tells me! And he hasn't got any children. And he isn't married or divorced. Wonderful, isn't it?
MAN	It is. And this is his office? Where is it? In the town centre?
WOMAN	No, no, not really. It's near the ring road. He's got offices and a warehouse there.
MAN	And where does he live? Near the ring road?
WOMAN	No, he lives in the country.
MAN	Very nice.
WOMAN	Yes, it is. He's got a lovely old house in a small village.
MAN	And how well do you know him?
WOMAN	Oh, not very well – not yet, anyway. I met him about a month ago. I needed a new back door so I phoned him and he came round.
MAN	Be careful, Janie. Don't mix business and pleasure!

Lesson 11 Listening, activity 1

Speaker 1 I'd really like to sell this old car and buy a new one. But first of all I need some money. I'm going to start saving now.

Speaker 2 I'd like to go and visit my grandchildren in Australia. I went there with my wife before she died, er... but I'd like to go again.

Speaker 3 I'd just like to have some time to myself. I love my children, don't get me wrong, but I never have any time when I can do what I want. I'm not going to stay at home forever. I'm going to look for an evening job.

Speaker 4 I really like playing football. And what I'd like to do is to play for the school team. I'm a good player – that's not the problem. But I'm not a boy.

Lesson 13 Vocabulary and listening, activity 2

SHEILA	I've got my traveller's cheques. They're in US dollars because that's the best for Latin America. I haven't got any Brazilian currency – but I can get that when we get to Rio. The exchange rate'll probably be better anyway. I'm not taking a handbag because I'll have this backpack on my back and then my traveller's cheques, money and passport in my wallet. Where's my penknife? Oh, good. I don't need scissors because there are scissors on the penknife. Right, then. I think I'm ready. I'm really excited!

Lesson 15 Listening, activity 1

WOMAN	How about chicken supreme, then? That looks quite nice.
MAN	OK. What do we need then, apart from chicken?

WOMAN	Er... an onion. Can you start a shopping list?
MAN	An onion, some chicken...
WOMAN	Yes, two pieces of chicken, some lemon juice – oh, put a lemon on the list, can you? Some mushrooms... oh no, we've got some mushrooms. An egg yolk... no, we've got eggs already...
MAN	What are we going to have with it? Potatoes?
WOMAN	No, let's have rice. But we haven't got any, so can you put a packet of rice on the list? And bread – a loaf of brown bread – put that on the list too. Oh, and we need some cream. Have we got any?
MAN	Yes, I bought some yesterday. I'll get a bottle of wine too. White?
WOMAN	Yes, that's nice with chicken. What about dessert? I know, I'll do peaches in red wine. We've got some red wine. Can you get two peaches – nice, ripe peaches, and a packet of sugar. We haven't got very much.
MAN	Sugar... OK. Is that everything?
WOMAN	Yes, I think so.

Lesson 16   Listening, activity 1

LOUISE	So, let's do something on Friday then. I've got the paper in front of me so let me see what's on. Mmm... there's the Kirov Ballet at the Playhouse.
FRIEND	I like ballet. That would be nice.
LOUISE	Oh, but it's their last two performances today, so we're too late for that. Pity! How about a musical?
FRIEND	What is it? I don't like musicals very much.
LOUISE	*Guys and Dolls.* It ends on Saturday.
FRIEND	No, I don't really want to see it,
LOUISE	Well, how about the cinema?
FRIEND	What's on?
LOUISE	Just a minute... er... *Jurassic Park.*
FRIEND	Oh, yes! I didn't see it when it first came out. Where's it on?
LOUISE	At the Showcase. You know, the cinema in Grey Street.
FRIEND	Mm... I'm just thinking... What time does it start?
LOUISE	Well, it's on at 2.10, 5.10 and 8.10.
FRIEND	Let's go to the 5.10 and then have something to eat afterwards.
LOUISE	All right, then.
FRIEND	OK. Well, let's meet outside the cinema at five o'clock.
LOUISE	OK. See you on Friday then.

Lesson 17   Listening, activity 1

Well, she's not very tall – quite short – and she's got dark hair.

Very short – she went to the hairdresser last week so it's very short at the moment.

Definitely like her mother – they're both the same build and they have very similar characters too – her mother's really kind too.

She's thirty. Thirty's a bit special so we had a big party for her thirtieth in February.

Lesson 17   Listening, activity 2

Well, he's really intelligent – I'm not sure that that's a good thing! Yes, intelligent and very calm. I'm the opposite.

He looks like his brother – they're obviously brothers. He doesn't really look like his mum or dad though.

Grey – well, he's going grey now that he's getting older. He's going grey and bald and he's not very happy about it!

He's average height for a man – medium-height, I'd say. Taller than me, anyway.

Lesson 18   Vocabulary and listening, activity 2

Speaker 1 My husband's always pessimistic – he always thinks the worst is going to happen. I don't like that about him, I don't like it at all. He's a sensible person, and that's a good thing, because I'm not. I'm not sensible at all, so I like that in him. Then he isn't very confident. I don't know why, but he isn't very confident

and there's no reason why not. This sometimes means that he isn't very friendly, which is a pity. People are only friendly to you if you're friendly to them.

Speaker 2 My boyfriend's honest and kind. I like both those things about him – his honesty and kindness. I'm not so kind, at any rate. So those are the things I admire about him. What I don't like is the fact that he isn't tidy at all – his things are always in a mess – and then he isn't very reliable. He's always late for everything and I don't like that because I'm usually the one who's waiting.

Speaker 3 My partner's an interesting character – that's why I like her. She's older than me and she's spent a lot of her life abroad, and that makes her interesting. She's also imaginative. I'm quite imaginative too so I like that about my partner. On the negative side, she isn't very calm – she gets angry easily, which I don't like. And she isn't very patient, and I don't like that either.

Lesson 21   Listening, activity 1

Speaker 1 Yes, I have. I cut my thumb quite badly last year when I was cleaning my penknife. Then about two years ago I fell and cut my arm when I was running in a cross-country race. That was probably worse.

Speaker 2 Well, not seriously ill. When I was in Equador eight years ago, I had an upset stomach. But it was my own fault. I drank the water, you see and you should only drink it after you've boiled it.

Speaker 3 No, I haven't. I smoke – probably about ten or twenty a day – but I'm lucky, I've never had a cough, let alone a dreadful cough.

Speaker 4 No, never. I like a glass of wine now and again but I don't think I drink too much. I'm quite slim, so I don't need to give up chocolate or anything sweet. Anyway, I don't eat many sweet things.

Lesson 22   Listening, activity 1

Speaker 1 Yes, I'm really pleased with myself. I've just passed my driving test. It was my third attempt – but at least I've passed it.

Speaker 2 What's new? Mmm, well, I've just started taking tennis lessons. I wanted to get fit, so I thought I'd like to take up tennis.

Speaker 3 I've just stopped smoking and it's killing me! I've become nervous – more nervous than before I stopped. and, of course, I'm eating more now.

Speaker 4 When we met last I lived in a flat with four other people. Well, now I've got my own flat. I've just moved in.

Lesson 23   Listening, activity 1

Speaker 1 I'm really boring, I'm afraid. I have never sent a Valentine's card. This doesn't mean I've never received any, mind you. I think I've probably had about six or seven all told. And once I got a beautiful bunch of flowers. I never found out who they were from.

Speaker 2 I've sent Valentine cards. One year, in fact, when I was at school, I sent five to the same girl, but, er... she moved away and I didn't see her again. How many I have received? Um, one, I think and that was last year from my wife. So I took her out for a meal. That's the only thing I've done on Valentine's Day – apart from sending cards.

Speaker 3 I sent quite a few when I was younger, and I got quite a few too. I have to admit that I got married on Valentine's Day – that was in 1987. My girlfriend – now my wife – and I didn't choose Valentine's Day, we just wanted to get married on the second Saturday in February, and that was Valentine's Day. I'll never forget it! We got lots of cards that year.

Lesson 24   Listening, activity 2

WOMAN	The place where I hang my clothes is called a 'closet'.
MAN	Oh, yes, we call that a 'cupboard'.
WOMAN	Oh.
MAN	Um, when you go into a building, like a house, for example, you walk in, and you're on the 'ground floor'.

WOMAN	Oh, no. In America, you'd be on the 'first floor'.
MAN	Uh huh.
WOMAN	If I don't have any money, and I want to call my parents, I 'call them collect', and they pay for it.
MAN	Oh, I see what you mean – we call that a 'reverse charge call', you 'make a reverse charge call'. You can get a vegetable in this country, which is purple and oval-shaped, and, er... we call it an 'aubergine'.
WOMAN	Oh, that's what an aubergine is! We call it an 'egg plant'.
MAN	Oh, yes.
WOMAN	If I'm in a tall building, I usually don't take the stairs – I take the 'elevator'.
MAN	Ah, um... we call that a 'lift'.
WOMAN	Oh.
MAN	Yeah. Um... if you go to the theatre or to the cinema, say, um... and there are a lot of people, you have to make a queue outside – you 'queue' outside.
WOMAN	We don't! We 'stand in line'.
MAN	Ah.
WOMAN	One of the things we might stand in line for, though, is a 'movie'.
MAN	Oh, in a cinema, yes, we call that a 'film' – you go to see a film. Um... when you... throw out all the things in your house, you know, empty cartons, stuff you want to throw away – we call that 'rubbish'. You put it out with the rubbish.
WOMAN	Oh, right. We call it 'garbage'.
MAN	Oh, yes.
WOMAN	If I'm travelling somewhere, and I don't intend to return, I'd buy a 'one-way ticket'.
MAN	A one-way ticket... oh, we call that a 'single ticket'.
WOMAN	A single?
MAN	Yeah. Um... on the other hand, if you're travelling and you want to get to the other place and then come back to the same place, we call that a 'return ticket'.
WOMAN	Oh, we call it a 'round-trip ticket'.
MAN	Oh, right.
WOMAN	If I was moving house and had a lot of things that wouldn't fit in my car, I'd probably take them in a 'truck'.
MAN	A truck... we call that a 'lorry'.
WOMAN	Oh.
MAN	Yeah, you put them in a lorry. Um... if you go to the shops you can buy children lovely – and grown-ups for that matter – they love 'sweets', or chocolate.
WOMAN	Mmm! We call that 'candy'...
MAN	Ah, right.
WOMAN	...and I love it!

Lesson 25   Vocabulary and listening, activity 2

1 It's oblong in shape. It's made of leather and metal. The leather's quite hard.
2 It's round and soft, and it's made of wool.
3 It's made of wood and glass. It's oval and flat.

Lesson 25   Reading and Listening, activity 2

Right, 'anorak'. Well, it's a kind of coat, a warm coat made of a material like nylon or PVC, which is waterproof and, er, warm – because you wear it in winter. It's fairly short, fairly floppy and it comes to the top of your legs, or thereabouts.

'Eiderdown'. Um, it's a rectangular thing that you put on your bed. Um... it's big, um... usually made of cotton, and it's full of feathers, to keep you warm, and you sleep underneath it and it gives you a really nice night's sleep.

'Spanner' – um... a spanner is a tool, um... which you use when you're working on a piece of machinery, like a car, for example. And it's made of metal and it's long and it's thin, and it's got two round bits at each end, um... which you use for doing up, or undoing nuts.

Lesson 25   Reading and listening, activity 4

'Dummy'. It's a thing that you give to a baby to stop it crying, normally when the baby's hungry. It's usually made of plastic, with a sort of rubber teat on the end, and you put it in the baby's mouth, and the baby sucks it.

'Dungarees'. An item of clothing, a pair of trousers – usually used for work – made of denim, or cotton, or some other material, which also has a bib part which comes up to cover the chest. They have braces which come over the shoulder to hold them up.

'Face cloth'. It's a small piece of cloth, usually made of cotton. It's square in shape, and you use it to wash your face, so that you don't have to use your hands.

'Lawn mower'. Yes, this is made of metal, and it's a machine for cutting grass. Er... it has a little box at the back which collects the grass. Um... there are still a few that you can work by hand, but now they're usually driven either by electricity or petrol.

'Oven cleaner'. Well, oven cleaner is a kind of foam which you use to clean the grime off your oven with. It's thick, white and sprays on the inside of your oven. You leave it for a while and, er... then you wipe it off with a wet cloth or a sponge.

'Trunk'. If you're moving house or you need to store your things somewhere, this is what you would use. Um... it's a large oblong-shaped box, um... often made of metal, um... and you put your things in it – clothes, books, um... objects, anything really. And, um... it's a very useful item.

Lesson 28   Vocabulary and listening, activity 2

Conversation 1
WARDEN	Now then, this is the kitchen. You can cook your own meals here if you want to.
BOY	Yes, we've brought some food with us. We'll use the kitchen.
WARDEN	Or you can have a meal in the restaurant.
BOY	I didn't know that youth hostels had restaurants.
WARDEN	Oh, yes, most of them have got a restaurant and they've all got shops. Well, you can buy most things you'll need in the shop.
BOY	Well, the next time I go youth hostelling I'm not going to carry my food with me!

Conversation 2
LIBRARIAN	Here's your ticket then. Please bring it with you when you want to take out some books.
WOMAN	Er... how many books can I take out at a time?
LIBRARIAN	You can take out five books at a time. But with this ticket you can take out five books from this library and five from the central library in town.
WOMAN	So, ten in total.
LIBRARIAN	That's correct.

Conversation 3
AIR STEWARD	Would you like anything to drink, madam? We've got orange juice, cola...
WOMAN	What about alcoholic drinks? Wine or whisky?
AIR STEWARD	You can have an alcoholic drink, madam, but you have to pay for it.
WOMAN	The last time I was on an aeroplane I didn't have to pay for an alcoholic drink.
AIR STEWARD	It was probably a different airline, madam. By the way, madam, this is the no-smoking area. You can't smoke here. I'll see if there's an empty seat in the smoking area.

Conversation 4
CUSTODIAN	Is that a camera you've got there, sir? I'm afraid you can't take photos of the exhibits.
MAN	I see.
CUSTODIAN	Leave it here in the cloakroom with us. Hand this ticket in when you come back and we'll return your camera.
MAN	What about my coat? Can I leave that too?
CUSTODIAN	Yes, of course, sir.

Lesson 29   Listening, activity 1

Conversation 1
MAN	What's the matter, Pat? You're not looking too good.
PAT	I've got a dreadful headache.
MAN	There's some aspirin in the bathroom cupboard. Why don't you take two of those?
PAT	I think I will. Thanks.

Conversation 2

WOMAN What's the best way to deal with jet lag? It's my bedtime now, English time, but it's only 6pm here in New York.

MAN I think you should try and forget that it's your normal bedtime.

WOMAN But I feel awful.

MAN The best thing to do is to try and fit into the time of the place where you are, as soon as you can.

Conversation 3

MOTHER What's the matter, Joanne?

JOANNE I've got a sore throat. It hurts.

MOTHER Oh, dear. Why don't you have a hot lemon drink? I'll make you some.

JOANNE Thanks.

Conversation 4

MAN Mrs Jenkins. I'm sorry about this, but I think I've got flu. I think I'll have to go home.

MRS JENKINS All right. You do that. We don't want other people in the office catching it from you, do we?

MAN No.

MRS JENKINS Go home and go to bed for the rest of the day. See how you feel in the morning.

Lesson 30   Listening, activity 1

Conversation 1

STUDENT Do you think you could spell that, please?

TEACHER What? 'Sympathetic'? S-Y-M-P-A-T-H-E-T-I-C.

STUDENT Sorry, what's after the 'p'?

TEACHER A-T-H-E-T-I-C. Sympathetic.

STUDENT Thank you.

Conversation 2

STUDENT Would you mind if I left early today? I have to go to the dentist's, you see.

TEACHER The dentist's! Oh, you poor thing! What time's your appointment?

STUDENT Er... three o'clock.

TEACHER And what time do you want to leave?

STUDENT 2.30, if that's OK.

TEACHER Yes, that's fine.

Conversation 3

STUDENT Can we do this activity in pairs?

TEACHER Yes, if you want. Work with your neighbours – but there's no need to shout!

Conversation 4

STUDENT Can we listen to that tape again?

TEACHER Who wants to listen to the tape again?

STUDENTS Mm... yes... yes...

TEACHER All right, let's listen one more time. You can look at the tapescript at the back of the book, if that will help.

Lesson 33   Listening, activity 1

Speaker 1 I think there should be schemes which encourage people to share their cars. I mean I think you shouldn't be able to drive a car without a passenger. I think there's a highway near Washington DC in the States – I think I've been on it actually – and there must be more than one person in each car or you're fined. I think that's quite a good idea.

Speaker 2 In winter I think we should wear more clothes. Well, most people have their central heating on very high, but I think we could save energy by wearing more clothes.

Speaker 3 I think we should all adopt a tree, you know, we should take more interest in the growing of trees. One of my aunts gave me this idea – she supports a tree-planting scheme in her area.

Speaker 4 We mustn't buy aerosols which aren't CFC-free. CFCs – chlorofluorocarbons – are very damaging to the ozone layer and to the environment, so we must make sure that aerosols are free of them. I think most of them are these days.

Lesson 36   Listening, activity 2

FORECASTER And now the outlook for the weekend for London and the south-east. Tomorrow will be dry and sunny, with the occasional shower on the east coast. Temperatures will reach about 22° centigrade, though there may be a light wind. Things are not looking so good for Sunday, however. Much of the day will be cloudy, with quite a strong wind for this time of year. Rain is possible in all parts of the south-east in the afternoon, though showers will die out towards the evening. Maximum temperature 17° centigrade. That's all for now. Have a good weekend.

Lesson 37   Listening, activity 2

FIRE OFFICER The most important thing to remember if you can't get out of your home because of a fire is to keep calm. If you keep calm, you can save your energy – and you can use this energy to help you survive. If you're prevented from getting out of your home or a building because of flames or smoke, close the door – the door nearest to the fire – and use towels or sheets to block any gaps. If you put towels at the bottom of the door, this will help stop smoke spreading into the room. After that, go to the window. Get down to floor level if the room becomes smoky. It's easier to breathe down there because the smoke will rise upwards. When you get to the window, try to open it and to call out to people outside. They can call the fire brigade. You should just wait for the fire brigade – they should arrive in a short time. On the other hand, if you're in immediate danger, you should think about getting out of the window. So drop cushions or bedding – if they're in the room with you – to the ground to break your fall from the window. Get out of the window with your feet first and lower yourself to the full length of your arms before dropping.

Lesson 38   Listening, activity 1

Speaker 1 I'd go with one other person – someone I know quite well. The holiday would be a mixture of sun and beach and adventure, walking, visiting places. And the food would be something special.

Speaker 2 I'd have a big garden, and a gardener to look after it. I'd like a house in the country with lots of rooms. And then when the rooms got untidy, I could close the door and move into the next room.

Speaker 3 It would never, ever break down. And it wouldn't use much petrol so I could travel a long way and not worry about looking out for petrol stations.

Speaker 4 This would be big and light and fresh. And furniture... mmm... I think there's only be my desk and two chairs. There wouldn't be any cupboards – my secretary would look after everything!

Lesson 40   Listening, activity 2

MARY Well, I was travelling across France after a skiing holiday – travelling back to Britain. Er... we were in Tony's car. Tony was driving and there was another friend, another person with us, John. We decided not to go on the motorways but to go on smaller roads because we didn't want to pay the motorway tolls. We wanted to save the money instead. Anyway, we were driving along, and unfortunately, because the main road was clear – there was no traffic coming – Tony pulled out of a side road without stopping. Just round the corner, of course, there was this policeman who fined Tony 1,000 francs – that was more than £100 because he hadn't stopped at the 'stop' sign at the road junction. Because the three of us, we had all wanted to save money, I thought we should divide the 1,000 francs between the three of us. I mean I didn't think Tony should pay the lot because he was driving on smaller roads so that we'd all save money. Anyway, this fellow John, refused to pay anything towards the fine. He and I had this big argument about the money. It was pretty awful because I don't usually argue with people. In the end Tony and I divided the fine between us, and, I have to say, I haven't spoken to John since. Not once.

Answer Key

Lesson 1

READING
2 'What shall I call you?'

SOUNDS
1 1 <u>Pleased</u> <u>meet</u>   2 How <u>old</u>
3 How <u>do</u>   4 What's <u>first</u> <u>name</u>
5 <u>help</u>

VOCABULARY
1 Across: 1 help   4 where   5
family   7 are   8 meet   10 first
11 do   13 understand
Down: 1 hello   2 please
3 repeat   5 from   6 married
9 is   old

2 1 e   2 c   3 a   4 b   5 d

3 Across: want   sing   go   visit
live   talk
Down: arrive   think   stay
take   offer   drink   ask   sit
accept

GRAMMAR
1 1 visit   2 take   3 talk   4 ask
5 drink   6 go
3 1 I never offer to wash the
dishes.
2 I always arrive about ten
minutes late.
3 I usually take wine or
chocolates.
4 I sometimes want coffee
after the meal.
5 I don't often go to dinner
parties on my own.
6 I don't usually wear smart
clothes.
5 *Example answers*
you Are you American?
you Where do you live?
you What do you do?
you Are you married?
you What's your name?

Lesson 2

VOCABULARY
1 1 seven pm / seven o'clock in
the evening
2 one fifteen am / a quarter
past one in the morning
3 two forty-five am / a quarter
to three in the afternoon
4 nine thirty am / half past
nine in the morning
5 eleven forty-five at night / a
quarter to twelve at night
6 eight ten am / ten past eight
in the morning
2 *Example answers*
1 I wake up   2 I get up   3 I
have breakfast   4 I get ready
for school/work   5 I leave

home   6 I have lunch   7 I
come home   8 I have dinner
9 I watch TV   10 I go to sleep

SOUNDS
1 1 comes   goes   has
2 gets   stops   works
3 dances   finishes   washes
4 asks   sits   takes
5 dresses   refuses   watches
6 lives   offers   serves

GRAMMAR
1 1 goes   2 works   3 finish
4 sits   5 doesn't   6 watch
7 offer   8 don't
3 1 up   2 home   3 lunch
4 and   5 doesn't   6 day
4 1 gets   2 has   3 washes
4 watches   5 gets   6 goes

LISTENING AND WRITING
1 The person in photo 3 is
speaking.
2 8.30am They leave home for
school.
12.30pm She stops for lunch.
6.00pm They have dinner.
10.30pm She goes to bed.
3 Read the tapescript and check
your answers.

Lesson 3

GRAMMAR
1 1 -   2 a   3 the   4 -   5 a   6 -
7 the   8 an   9 -   10 the
11 an   12 a   13 -   14 an
15 the   16 a   17 the   18 a
19 the   20 the
4 1 buses   2 curtains   3 classes
4 families   5 parties   6 plays
7 sandwiches   8 shoes
9 windows   10 women

SOUNDS
1 1 <u>house</u> <u>flat</u>   2 <u>main</u> <u>home</u>
3 <u>bedrooms</u>   4 <u>dishwasher</u>
5 <u>eat</u> <u>meals</u>   6 <u>television</u>

LISTENING AND WRITING
1 1 flat   2 living room   3 two
4 yes   5 living room   6 yes
2 Read the tapescript and check
your answers.

READING
1 1 region or country where a
person was born   2 a house
where people live and are
looked after   3 the place
where a person lives   4 a
game on the team's own
ground   5 not very special
6 find a place to keep things
7 confident and able   8 a
place where someone feels
very happy and relaxed

Lesson 4

READING AND WRITING
1 1 radio   2 parks   3 football
4 queues   5 winter
6 newspapers   7 gardening

SOUNDS
1 Words with two syllables:
awful   crowded   dirty
driving   friendly   hotels
litter   police   polite
shopping   tourists   weather
2 Sentences 1,2, 4 and 5 use a
strong intonation.

VOCABULARY
1 1 crowded   2 driving   3 food
4 friendly   5 litter   6 cheap

GRAMMAR
2 *Example answers*
1 visiting   2 going to   3 having
4 listening to   5 giving
6 going   7 drinking   8 playing

Lesson 5

SOUNDS
1 1 <u>waiting</u> <u>someone</u>   2 <u>doing</u>
<u>shopping</u>   3 <u>standing</u> <u>road</u>
4 <u>playing</u> <u>accordion</u>
2 1 <u>painting</u> <u>picture</u>   2 <u>walking</u>
<u>work</u>   3 <u>thinking</u> <u>friend</u>
4 <u>holding</u> <u>something</u>

GRAMMAR
1 ask   2 carry   3 go   4 watch
5 come   6 leave   7 shine
8 write   9 get   10 sit   11 stop
12 travel
2 1 wears / is wearing   2 works
/ is working   3 drink / am
drinking   4 am having / have
5 watches / is watching
6 read / am reading

VOCABULARY
1 1 N   2 V   3 V   4 N   5 N   6 N

READING AND WRITING
1 1 a   5 b   10 d   12 c
2 e A woman is talking on the
phone.
f A man is writing a letter.
g A man and a woman are
dancing.
h A girl is riding a bicycle.

Lesson 6

SOUNDS
1 1 expected   started   visited
2 finished   liked   walked
3 enjoyed   lived   tried
4 danced   stopped   watched
5 continued   stayed   travelled
6 listened   opened   replied

GRAMMAR
1 1 started   2 walked   3 lived
4 watched   5 asked   6 opened
2 1 didn't play   2 didn't like
3 didn't look   4 didn't dance
5 didn't stay   6 didn't want
3 left   said   gave   told   wrote
did   got   thought   came   knew
become   have   meet   sleep
hear   run   sit   take   be   make
4 became   had   met   slept   heard
ran   sat   took   was/were   made
leave   say   give   tell   write
do   get   think   come   know
5 1 PA   2 PR   3 PA   4 PR   5
PA   6 PR

VOCABULARY
1 1 Smart people   2 blue flower
3 shoes / flat   4 bag / heavy
5 old man   6 low / cloud

LISTENING
1 3 4 2 6 5 1

2 Brighton -   Criccieth 4   Deal 2
Hastings 3   Liverpool -
London 1   Margate -   Southend -
St Andrews 6   Stranraer 5

Lesson 7

SOUNDS
1 The speaker sounds interested
in sentences 1, 3, 4 and 6.

VOCABULARY
1 1 I took *a taxi* to the airport.
2 I changed <u>my traveller's</u>
<u>cheques</u> at the bank.
3 She carried <u>cash</u> in her
handbag.
4 We didn't choose <u>a package</u>
<u>holiday</u> last year.
5 A young receptionist
checked <u>the coupons</u>.
6 Our holiday guide met <u>us</u> at
the airport.
2 See the underlined words above.

LISTENING AND WRITING
1 Speaker 1: When was your
holiday?
Speaker 2: Who did you go with?
Speaker 3: Where did you go?
Speaker 4: Which part of the
holiday did you enjoy most?
2 Speaker 1: January this year.
Speaker 2: With friends.
Speaker 3: South-west Turkey.
Speaker 4: Going down the
Grand Canyon.
3 Speaker 1: Question 1
Speaker 2: Question 5
Speaker 3: Question 4
Speaker 4: Question 6

GRAMMAR
3 1 When did The Titanic hit an
iceberg and sink?
2 Who went to the moon in
1969?
3 What did Charles Lindbergh
do in 1927?
4 Where did Marco Polo go?
5 Who went to the South Pole
in 1912?
6 When did Sir Edmund Hillary
and Sherpa Tenzing reach the
top of Everest?
7 What opened to the public
in 1955?
8 What did Jules Verne write?

Lesson 8

SOUNDS
1 1 thought   2 got   3 left
4 gave   5 hit

GRAMMAR
1 1 got   2 had   3 got   4 went
5 had/ate   6 wrote   7 went
8 had   9 took   10 saw

11 was 12 came
13 made/cooked 14 had/found
15 had 16 left/went 17 was
2 *Example answers*
1 He didn't stay in bed late.
2 He didn't have a bath.
3 He didn't drive to the post box.
4 He didn't have lunch with his brother and his family.
5 He didn't take his nephews to the theatre.
6 He didn't go out for a meal in the evening.
4 1 b, e 2 a, d 3 c, f

READING AND WRITING
1, 3, 5, 4, 1, 2

Lesson 9
READING
2 1 N 2 N 3 N 4 N 5 N
3 2 Kathy doesn't actually give the answers to any of the questions.
3 The text gives clues to some of the answers: 1 don't know 2 probably five: Kathy, her husband and three children 3 probably three 4 don't know 5 Greek-Australian, but probably more Greek than Australian

VOCABULARY

♀ aunt girlfriend daughter mother girl grandmother wife niece sister woman

♂ uncle boyfriend son father boy grandfather husband nephew brother man

SOUNDS
1 1 children 2 haven't sisters
3 parents 4 from a 5 haven't cousins

GRAMMAR
3 1 He is my father.
2 We are your brothers.
3 I am her cousin.
4 They are our girlfriends.
5 She is his wife.
6 You are their nephews.
4 1 S 2 P 3 P 4 S 5 P 6 P
7 S 8 S
5 1 sister's 2 brother's 3 cousins'
4 friends' 5 sisters' 6 nephew's

LISTENING
1 *Example answers*
1 Who is he?
2 Where does he live?
3 How old is he?
4 What does he do?
5 Is he married?
6 How well do you know him?
2 The speaker answers all the questions.
3 1 James Smith. 2 In a lovely old house in a small village in the country. 3 32. 4 He sells doors and windows. 5 No.
6 Not very well.

Lesson 10
SOUNDS
1 1, 2, and 5.

VOCABULARY AND READING
2 1 Kalo Chorio 2 Aghios Nikolaos 3 Kalo Chorio
4 Kalo Chorio
3 1 It's got three.
2 Yes, they have.
3 It's got one post office.
4 Kalo Chorio hasn't got a market.

GRAMMAR
1 1 How many supermarkets has Aghios Nikolaos got?
2 Has Aghios Nikolaos got a doctor?
3 Has Kalo Chorio got any beaches?
4 Has Aghios Nikolaos got a post office?
2 1 has 2 have 3 has 4 have
5 have 6 has
3 2 We've 3 Madrid's 5 They've 6 It's

WRITING
1 1 The architecture is interesting but the streets are dirty.
2 The public transport is safe and clean.
3 It's got excellent but expensive restaurants.
4 It's crowded and dangerous.
5 It's got a couple of parks but it hasn't got any swimming pools.
6 The shops are cheap but crowded.
2 Here we are in Kalo Chorio. We're having a lovely time (1) and the weather's great. Martin's nice and brown (6) but I'm just red. We've got a nice apartment (5) and it's only five minutes' walk from the shops and tavernas. There isn't much to do here (2) but it's very relaxing. There's only one disco (4) but there are quite a lot of bars. We're sitting in Pinnochio's (3) and having a drink and a barbecue. Hope all's well at home. See you soon.
Love, Diane x x

Lesson 11
GRAMMAR
1 1 ...so I'm going to start saving some money
2 ...so I'm going to teach myself a foreign language.
3 ...so I'm going to do some serious training.
4 ...so I'm not going to eat junk food.
5 ...so I'm going to phone the airport and see if you can have lessons.
6 ... so I'm going to look at the

ads in tomorrow's paper.
2 1 I'm going to start saving some money because I'd like to buy a horse.
2 I'm going to teach myself a foreign language because I'd like to work abroad.
3 I'm going to do some serious training because I'd like to run a marathon.
4 I'm not going to eat junk food because I'd like to lose some weight.
5 I'm going to phone the airport and see if you can have lessons because I'd like to try parachuting.
6 I'm going to look at the ads in tomorrow's paper because I'd like to change my job.
4 *Example answers*
1 Are you going to learn to drive?
2 Are you going to get up late this weekend?
3 What are you going to do?
4 Are you going to catch a bus?
5 Are you going to watch it?
6 When are you going to tidy it up?
5 1 going 2 to have 3 being
4 to change 5 to live 6 doing

WRITING
1 She's going to look after some children because she'd like to improve her German, and the children are German.

SOUNDS
1 last school abroad first qualifications place Leeds University going economics Languages important business summer look after children Germany improve French

LISTENING
1 Speaker 1: buy a new car, start saving
Speaker 2: visit his grandchildren
Speaker 3: have some time to herself, look for an evening job
Speaker 4: play for the school team
2 Read the tapescript and check your answers.

Lesson 12
READING
1 2 Language teaching in British schools

SOUNDS
1 1 She won't take her shoes off.
2 It'll be there at six o'clock.
3 We don't want a large meal.
4 He likes discos a lot.
5 I'll stop smoking soon.
6 I go to school on my bike.

VOCABULARY
1 *Example answers*
accountant: arithmetic, maths
doctor: biology, chemistry, physics, science
engineer: arithmetic, chemistry, maths, physics
journalist: economics, geography, history, languages
politician: economics, geography, history, languages
secretary: languages

GRAMMAR
1 1 I'll 3 They'll 4 I'll
2 1 I think Susan will become rich and famous.
2 I think Peter will decide to leave home.
3 I think Katie will get married in September.
4 I think William will learn another foreign language.
5 I think Joanna will make some new friends.
6 I think Eddie will start his own business.

Lesson 13
GRAMMAR
1 1 I'll 2 She's going to 3 I'll
4 are you going to/We're going to 5 We'll 6 He'll
2 1 this evening 2 tomorrow morning 3 the day after tomorrow 4 in three days' time 5 next month 6 in a year's time
4 1 'll 2 won't 3 won't 4 'll
5 won't 6 'll
5 1 I won't eat so much fast food.
2 I'll be early for lessons.
3 I'll be less pessimistic.
4 I won't drive to work.
5 I'll go to bed earlier.
6 I won't watch so much television.
6 1 won't 2 'll 3 won't 4 'll
5 'll 6 won't
8 1 going to 2 will 3 going to
4 will
9 1 I'm going to visit my grandparents at the weekend.
2 I'll make you a drink.
3 My dad's going to phone at six o'clock.
4 I'll go there on Saturday morning.

SOUNDS
1 aspirin backpack camera handbag map passport sleeping bag travellers' cheques walkman

VOCABULARY AND LISTENING
1 backpack, camera, guide book, handbag, medical kit, pair of scissors, passport, penknife, toothbrush, toothpaste, traveller's cheques, wallet, watch
4 handbag, pair of scissors
5 Because she'll have her backpack and her wallet, and scissors on the penknife.

Lesson 14

READING

3 1 George's cafe
 2 Queen's Lane Coffee House
 3 The Randolph Hotel
 4 The Wykeham Coffee Shop
 5 The Nosebag

4 eat: cucumber sandwiches, crumpets, scones with jam and cream, Dundee cake, homemade cakes drink: Earl Grey tea, tea

VOCABULARY

1 bus station, Christ Church College, New College, Bodleian Library, Oxford Town Hall

SOUNDS

1 colleges houses offices
2 factories libraries universities
3 hotels hospitals stations
4 banks restaurants shops

GRAMMAR

1 *Example answers*
The Town Hall is in St Aldate's next to Christ Church College. New College is in Holywell Street, opposite the Wykeham Coffee Shop.
The Bodleian Library is in Radcliffe Square, on the corner of Brasenose Street and Catte Street.
The Covered Market is between Market Street and High Street.

2 in front left next over end right left on left
3 *Example answers*
1 Go along St Aldate's to the crossroads and turn right into the High Street. Cross the road and turn left into Turl Street. At the end of Turl Street turn right into Broad Street. The Sheldonian Theatre is on your right, next to the Bodleian Library.
2 Go to the end of George Street. At the crossroads, cross over into Broad Street. At the end of Broad Street turn right into Catte Street. Radcliffe Square is in front of you, past the Bodleian Library.
3 Walk to the end of Holywell Street, and turn left into Catte Street. Cross Radcliffe Square and walk down Brasenose Street. At the end, cross Turl Street and the Covered Market is on the corner of Turl Street and Market Street.

Lesson 15

VOCABULARY

1 1 lettuce 2 fish 3 apple
 4 vegetable 5 juice 6 bread
2 1 apple, onion, orange, egg
 2 beer, water, juice, milk, wine
 3 coffee, butter, cheese, tea
 4 coffee, tea 5 beef, chicken, lamb, apple, carrot, potato,

banana, orange, peach, cheese, egg, bread 6 beef, chicken, lamb, apple, carrot, potato, banana, orange, peach, cheese, egg, bread

3 1 an egg, some eggs, some egg
 2 some bananas, some banana, a banana
 3 some chickens, a chicken, some chicken
 4 an onion, some onion, some onions

LISTENING

1 lemon juice, egg, onion, bread, chicken, cream, rice, sugar, wine, peaches, mushrooms
2 an onion, two pieces of chicken, a lemon, a packet of rice, a loaf of bread, a bottle of white wine, two peaches, a packet of sugar

GRAMMAR

1 1 I'd like 2 He's having
 3 I'm eating 4 I like
 5 Would you 6 We drink
2 1 How many shall we get?
 2 How much do we need?
 3 How many would you like?
 4 How much have you got?
 5 How much shall we get?
3 1 some. We don't need any apples.
 2 any. I've got some lettuce.
 3 any. He needs some eggs.
 4 some. She hasn't got any fish.
 5 some. I don't need any milk.

Lesson 16

GRAMMAR

1 at: half past five, night, three o'clock, the weekend
 in: August, 1994, the morning, winter
 on: Friday evening, 2nd November, Sunday 16 May, Wednesday
3 1 Let's go shopping on Saturday morning.
 2 How about going to the theatre in the afternoon?
 3 Why don't we have dinner in the evening?
 4 Would you like to try the new French restaurant?
6 1 I 2 I'd 3 I 4 I 5 I'd 6 I'd

SOUNDS

2 1 I'd like 2 We'd love
 3 They like 4 I'd love
 5 We like 6 They love

READING

1 1 film 2 musical
 3 exhibition 4 ballet
2 1 F 2 T 3 F 4 F 5 T 6 F
 7 F 8 F

LISTENING

2 What's on? Where's it on?
 What time does it start?
3 Showcase Cinema, Grey Street. Meet outside the cinema at 5 o'clock.

Lesson 17

VOCABULARY

1 appearance: attractive, good-looking, pretty, slim, tall, ugly character: calm, confident, intelligent, nervous, quiet, thoughtful
2 1 She's got long hair. It's fair and straight. She's tall and thin.
 2 He's got a moustache and a beard. He's bald. He's short and fat.
 3 She's medium-height and middle-aged. She's got dark, curly hair and glasses.
3 He's got dark hair and a beard. He's tall, slim and good-looking.
4 *Example answer*
She's short and slim. She's got white hair and glasses.

SOUNDS

1 a long's hair b old c Who's like d What look e Who look f What's like g tall h colour's hair

LISTENING

1 1 d 2 a 3 c 4 b
2 1 f 2 e 3 h 4 g

GRAMMAR

2 1 How tall is she/he?
 2 How old is she/he?
 3 How long's her/his hair?
 4 What does she/he look like?
 5 Who does she look like?
 6 What colour's his hair?
 7 Who's she like?
 8 What's he like?

Lesson 18

READING

1 The writer is an American.
2 friendly emotional loud polite informal

VOCABULARY AND LISTENING

2 Speaker 1: Her husband is pessimistic and sensible. He isn't confident or very friendly.
Speaker 2: Her boyfriend is honest and kind. He isn't tidy or reliable.
Speaker 3: His partner is interesting and imaginative. She isn't calm or patient.
3 Read the tapescript and check your answers.

GRAMMAR

1 adjective	comparative	superlative
1 bad	worse	worst
2 tall	taller	tallest
3 close	closer	closest
4 good	better	best
5 friendly	friendlier	friendliest
6 kind	kinder	kindest
7 nervous	more nervous	most nervous
8 nice	nicer	nicest
9 sensitive	more sensitive	most sensitive
10 smart	smarter	smartest

2 1 bad 2 taller 3 close
 4 best 5 friendlier 6 kind
 7 most nervous 8 nicest
 9 more sensitive 10 smart
3 1 Who is the nicest?
 2 Who is the laziest?
 3 Who is the youngest?
 4 Who is the shortest?
 5 Who is the most pessimistic?
 6 Who is the most good-looking?
4 1 Bob 2 Carl 3 Bob 4 Alan
 5 Bob 6 Alan

Lesson 19

VOCABULARY

1 a blouse W a coat - a dress - a hat - a jacket W jeans B a shirt M shoes W M a skirt W socks M B a suit M a sweater - a swimsuit G a tie M tights W trainers B a T-shirt B

4 1 I always wear white socks.
 2 My sister sometimes wears jeans.
 3 She often wears black tights.
 4 Her brother never wears a hat.
 5 My children often wear red T-shirts.
 6 I never wear formal clothes.

SOUNDS

1 1 socks 2 jeans 3 sweater
 4 tights 5 blouse 6 trainers

GRAMMAR

1 adjective	comparative
good	better
great	greater
late	later
big	bigger
lazy	lazier
formal	more formal

2 1 expensive 2 old 3 bad
 4 thoughtful 5 great 6 friendly
3 1 I'm less tidy than my best friend.
My best friend's less untidy than me.
My best friend's more tidy than me.
 2 I'm more pessimistic than my brother.
My brother's more optimistic than me.
My brother's less pessimistic than me.
 3 I'm less polite than my sister.
My sister's less impolite than me.
My sister's more polite than me.
4 1 John isn't as tall as his brother.
 2 His girlfriend isn't as old as he is.
 3 Bob isn't as popular as Carl.
 4 His father isn't as casual as he is. or He's not as smart as his father.
 5 He isn't as pessimistic as his father.
 6 His mother isn't as thin as he is.

Lesson 20

SOUNDS
1 The speaker sounds polite and friendly in questions 1, 3 and 5.

GRAMMAR
3 1 How fast can you go?
2 How far is it? / How far is London?
3 How much is petrol? / How much does petrol cost?
4 How much are hotel rooms? / How much do hotel rooms cost?
5 How long does it take (to Cader Idris)?
6 How long does it take by train (from Oxford to Birmingham)?
4 1 How fast   2 How old
3 How long   4 How tall
5 How big   6 How much
7 How often   8 How many

VOCABULARY
1 1 highway   2 set off   3 speed limit   4 desert   5 mile
6 police patrol   7 petrol
8 passenger

READING
1 It refers to her bicycle ride around the world.
2 Because she was not athletic or young. She wasn't a keen cyclist and she didn't have a bike. She had no idea how to mend a puncture and she hated camping, picnics and discomfort.

Lesson 21

SOUNDS
1 1 head   2 calf   3 knee
4 waist   5 thigh   6 throat

VOCABULARY
1 ankle back ear elbow finger fingernail foot forehead hair leg lip mouth neck shoulder stomach thumb toe tooth wrist
2 bald: head
big: nose head ear foot lip mouth toe tooth
brown: hair
fat: face finger stomach
long: arm face nose calf thigh body fingernail foot hair leg neck toe tooth
slim: calf thigh waist body finger foot leg
3 I've got two: arms, heels, eyes, calves, knees, thighs, ankles, ears, elbows, feet, lips, shoulders, thumbs, wrists.

GRAMMAR
1 1 No, I haven't. I left school when I was sixteen.
2 No, I haven't. I don't like acting.
3 Yes, and I've played too.
4 Yes, I had an appointment last week.

5 Yes, I have. But I'm divorced now.
6 No, I haven't. I prefer classical music.
2
infinitive	simple past	past participle
become	became	become
break	broke	broken
drink	drank	drunk
drive	drove	driven
eat	ate	eaten
pay	paid	paid
run	ran	run
say	said	said
speak	spoke	spoken
teach	taught	taught
wear	wore	worn

3 1 eaten   2 broken   3 drunk
4 spoken   5 run   6 driven
6 1 Did   2 Has   3 Have   4 Did
5 Did   6 Has

LISTENING
1 Speaker 1: Have you ever cut yourself badly?
Speaker 2: Have you ever been ill on holiday?
Speaker 3: Have you ever had a dreadful cough?
Speaker 4: Have you ever given up something because of your health?
2 Read the tapescript and check your answers.
3 1 Did you go to hospital?
2 Have you ever taken malaria tablets?
3 Are you ever ill?
4 Are you fit?

Lesson 22

GRAMMAR
1 1 's   2 've   3 's   4 've   5 've
6 've
2 1 've   2 -   3 's   4 -   5 've   6 've
3 feel give go hear know leave meet read see speak take think
4 felt heard left met read thought
5 1 died   2 had   3 have lost
4 have/known
5 have/visited/went   6 was
7 did/change   8 has stayed
6 *Example answers*
1 He's bought a flat.
2 He hasn't changed job.
3 He hasn't got married.
4 He hasn't moved to his new flat.
5 He's taken some exams.
6 He's visited Turkey.

LISTENING
1 Speaker 1: passed something
Speaker 2: started something
Speaker 3: stopped something
Speaker 4: moved somewhere
2 Read the tapescript and check your answers.

Lesson 23

SOUNDS
1 one syllable: flag king queen
two syllables: par*ade* pic*nic* *sol*diers
three syllables: *ce*lebrate *hol*iday *pres*ident

VOCABULARY
1 parade   2 soldiers   3 picnic

READING
1 and 2
anonymous cards: People send them today.
fertility festival: There was a Roman festival on 14th February called Lupercalia.
saint of lovers: He is St Valentine.
third century: This is when St Valentine was killed.
14th February: This is St Valentine's Day.
3 a: paragraph 2   b: paragraph 4
c: paragraph 1
d: paragraph 3   e: paragraph 5

LISTENING
1 They have all received Valentine cards.
2 Read the tapescript and check your answers.

Lesson 24

GRAMMAR
1 1 gives money to bank customers.
2 gives parking tickets when drivers break the rules.
3 sells newspapers and magazines.
4 is learning to drive.
5 you can put used glass items for recycling.
6 means 'no parking'.
2 The things you can see are numbers 2, 4 and 6.
3 1 which   2 which   3 who
4 which   5 which   6 who
4 1 who   2 which   3 which
4 where   5 who   6 where

SOUNDS
1 A   2 A   3 B   4 B   5 A   6 A

LISTENING
1 and 2
US	GB
closet	cupboard
first floor	ground floor
to call collect	to make a reverse charge call
eggplant	aubergine
elevator	lift
to stand in line	to queue
movie	film
garbage	rubbish
one-way ticket	single ticket
round-trip ticket	return ticket
truck	lorry
candy	sweets/chocolate

3 Read the tapescript and check your answers.

READING
1 The languages of America.

Lesson 25

VOCABULARY AND LISTENING
1 a: flat light round soft small wool
b: big flat hard heavy leather metal oblong
c: curved flat glass hard oval small wood
d: flat long metal plastic thin
2 1 b   2 a   3 c
4 It's made of metal and plastic. It's long, thin and flat.

SOUNDS
1 1 What does it look like?
2 Is it soft?
3 What's it made of?
4 Is it oblong?
5 Is it very big?
6 How much does it weigh?

GRAMMAR
2 1 knife   2 mirror   3 beret
4 suitcase
3 1 A knife is for cutting things.
2 To look at yourself, you use a mirror.
3 A beret is for keeping your head warm.
4 To carry things, you use a suitcase.
4 *Example answers*
1 A comb is for combing your hair.
2 A notepad is for writing notes on.
3 A phonecard is for making phone calls.
4 A towel is for drying yourself with.
5 A tent is for sleeping in.
6 A wallet is for carrying money in.
5 1 102   2 96, 97   3 100   4 55
5 112

READING AND LISTENING
1 and 2   1 b   2 a   3 a
3 dummy 3b   dungarees 2b
face cloth 2c   lawn mower 1c
oven cleaner 3c   pushchair - trunk 1a
5 *Example definition*
'Pushchair': a chair on wheels used for carrying babies or children who are too young to walk.

Lesson 26

SOUNDS
1 1 train   2 walk   3 out
4 head   5 throw   6 you

89

GRAMMAR

1 6 5 3 1 2 4

2 1 have to   2 have to   3 have to   4 have to   5 mustn't   6 mustn't

3 1 must   2 must   3 mustn't   4 must   5 mustn't   6 must

4 1 to   2 to   3 to   4 -   5 to   6 -

VOCABULARY AND READING

2 The words in the text are: clear, cross, footpath, look, pavement, road, traffic, walk

3 Picture 1: b   Picture 2: a and c   Picture 3: e   Picture 4: d

3 *Example answers*
1 You must walk behind one another and keep to the side of the road.
2 You mustn't let your children run into the road. You must walk between them and the traffic.
3 You must wait until the road is clear. You mustn't run.
4 You must wear or carry something white or light-coloured.

Lesson 27

VOCABULARY

1 1 locked   2 climbed   3 shouted   4 waved   5 whispered

3 Across: drink use stand play draw ride climb read make speak see
Down: touch run drive sleep

GRAMMAR

1 1 can/can't stand   2 could/couldn't speak   3 can/can't drive   4 could/couldn't draw   5 can/can't drink   6 could/couldn't climb   7 can/can't see   8 could/couldn't play   9 can/can't touch

3 1 can't   2 can   3 can't   4 can   5 can't   6 can't

SOUNDS

1 1 b   2 c   3 b   4 a   5 b   6 c

Lesson 28

READING AND WRITING

1 1F 2T 3F 4T 5F 6T 7T 8F 9F 10F 11T 12F 13F 14F 15T 16T 17T 18T 19T 20F 21T 22F 23T 24F 25T

VOCABULARY AND LISTENING

1 1 in a youth hostel   2 in a library   3 in an aeroplane   4 in a museum

2 Read the tapescript and check your answers.

GRAMMAR

2 You can't eat or drink. You can cycle.
You can't drive here. You can't smoke or eat.
You can't sit here. You can park here.

Lesson 29

SOUNDS

1 headache sunstroke temperature thirsty tired

VOCABULARY

1 I'm: sick/thirsty/tired
My ear/leg/tooth hurts.
I've got a headache/sore throat/temperature.
I've got flu/jet lag/sunstroke.

LISTENING

1 Speaker 1: headache
Speaker 2: jet lag
Speaker 3: sore throat
Speaker 4: flu

2 1 take two aspirin
2 forget that it's your normal bed time
3 have a hot lemon drink
4 go to bed for the rest of the day

GRAMMAR

1 1 You should take two aspirin.
2 You ought to forget that it's your normal bed time.
3 You should have a hot lemon drink.
4 You ought to go to bed.

3 1 You shouldn't smoke.
2 You should take plenty of exercise.
3 You shouldn't go to bed late.
4 You should eat plenty of fresh fruit and vegetables.
5 You shouldn't work too hard.
6 You should learn to relax.

4 1 (✗)   2 (✓)   3 (✓)   4 (✗)

READING AND WRITING

2 1 The most popular sports in North America are baseball and football.
2 The spectators are Americans from every income and ethnic group.
3 The baseball season is spring and summer.
4 The football season is fall (autumn) and winter.
5 They are important because alumni (ex-students) give money to their old school; they give more if the college has a winning team.

Lesson 30

SOUNDS

1 c f a d g e b
The speakers sound polite and friendly.

GRAMMAR

1 1 (✓)   2 (✓)   3 (✗)   4 (✗)   5 (✓)

3 *Example answers*
1 I'll answer it.
2 Shall I get you some?
3 I'll make one.
4 Shall I do it for you?
5 I'll book them.
6 Shall I look for them?

4 *Example answers*
1 Could you open the door, please?
2 Can you turn the TV on?
3 Can I have that magazine?
4 Can I help you?

LISTENING

1 1 D   2 P   3 P   4 P

3 Read the tapescript and check your answers.

Lesson 31

READING

1 4 His father heard a ghost.

2 three o'clock: his mother and father twice heard Ben's ghost at three o'clock in the morning of December 14th

GRAMMAR

1 1 She was having breakfast at eight o'clock.
2 She was getting into her car at a quarter to nine.
3 She was having a cup of coffee at a quarter past eleven.
4 She was speaking on the phone at five o'clock.
5 She was playing tennis at a quarter past seven.
6 She was ironing her clothes at a quarter past eleven.

3 1 I was crossing the road when the car hit me.
2 I was watching TV when there was a power cut.
3 I was recording a film when the video broke down.
4 I was running in a race when I hurt on my ankle.
5 I was getting some money when the gunman came into the bank.
6 I was skiing without a hat when it started to snow.

VOCABULARY

1 closing time insect bite jet lag military service mobile phone musical instrument police officer roast chicken sleeping bag speed limit

2 *Example answers*
1 art gallery   2 boarding pass   3 car park   4 cheese sandwich   5 guidebook   6 railway line   7 single room   8 swimming pool   9 traffic lights   10 traveller's cheques

Lesson 32

SOUNDS

1 disappear discover favourite prison shiver

2 One syllable: grounds strange warn
Two syllables: evil prison shiver
Three syllables: disappear discover favourite

VOCABULARY

1 nouns: grounds prison
adjectives: evil favourite strange
verbs: disappear discover shiver warn

2 1 discover   2 grounds   3 favourite   4 prison   5 shiver   6 warn

GRAMMAR

1 1 While we were standing outside the church, it started to rain.
2 While the policewoman was taking notes, the thieves ran away.
3 While their mother was working, the children stayed at school.
4 While I was having lunch, my cousins arrived.
5 While he was having a shower, the telephone rang.
6 While we were climbing the mountain, my cousin broke his arm.

2 1 While they were travelling in France, they heard the news.
2 She came to my office while I was speaking on the phone.
3 While he was playing table tennis, he got sunstroke.
4 He scored his first goal while he was playing against his old team.

3 1 We put up our umbrellas and ran to the car.
2 She ran after them and caught one of them. The other one got away.
3 She collected them at about five o'clock when she left the office.
4 I asked them if they would like something to eat.
5 It stopped ringing before he could get to it.
6 A helicopter came and took him to hospital.

5 1 I got home   2 her father died   3 I sat down   4 I went to the museum   5 the lights went out   6 she arrived

8 1 was working   2 wanted   3 remembered   4 knew   5 packed   6 prepared   7 drew   8 was working   9 stopped   10 saw   11 was painting   12 said

Lesson 33

VOCABULARY
1 1 clean 2 wet 3 more 4 flat
5 rich 6 noisy

GRAMMAR
1 1 hot 2 hilly 3 industrial
4 peaceful 5 rainy
2 1 hot 2 hills 3 industry
4 peaceful 5 rain
3 1 There isn't enough heat in
our house in winter.
2 It's too hilly for cyclists.
3 The town isn't industrial
enough.
4 There isn't enough peace for
the baby to sleep.
5 It's too rainy to go out for a
walk.
4 1 There aren't enough
2 There's too much 3 There
are too many 4 There's too
much 5 There isn't enough
6 There are too many
6 1 There are fewer theatres
than cinemas in Britain.
2 There is less industry in
Northern Ireland than in
Wales.
3 There is less unemployment
in the south of England than in
the north.
4 There were fewer motorways
twenty years ago than there
are today.
5 There is less rain in summer
than in winter.

SOUNDS
industrial south Wales people
north poorer rural peaceful
north mountains north-west
hilly centre quite farmland
Wales beautiful

READING
1 1 T 2 C 3 T 4 C 5 T 6 T
7 C 8 C

LISTENING
1 Speaker 1: T Speaker 2: C
Speaker 3: C Speaker 4: C
2 Read the tapescript and check
your answers.

Lesson 34

GRAMMAR
1 build - built eat - eaten give
- given hold - held know -
known write - written
2 1 is known 2 is called
is celebrated 4 are eaten
5 are held
3 1 are decorated 2 are closed
3 eat 4 are sold 5 exchange
6 are given
4 1 In Britain Easter buns are
called 'hot cross buns'.
2 Hot cross buns are eaten on
Good Friday.
3 The buns are made with
dried fruit.
4 The buns are marked with a
cross before they are baked.

5 The buns are toasted before
they are eaten.
6 The buns are served with
butter.

Lesson 35

GRAMMAR
1 1 although I use it in puddings.
2 we eat out about once a week.
3 but we usually drink from
mugs.
4 although we usually have it
as a main course.
5 but I always have a good
dinner.
6 although I quite like toast
occasionally.
2 1 Although I like chocolate a
lot, I try to have only one
piece a day.
2 I have a sweet tooth but I try
not to have too much sugar in
my tea.
3 I like salad with meat or fish
although my husband prefers
cooked vegetables.
4 We have a light breakfast
during the week. However, we
have a large brunch on Sundays.
5 We quite like pasta but we
don't have it very often.

SOUNDS
1 1 dish 2 saucer 3 slice 4
steak 5 spoon

VOCABULARY
1 fork glasses knife napkin
plates spoon table cloth
2 the woman: fish, salad, bread
the man: steak, egg, chips,
peas, bread

Lesson 36

GRAMMAR
1 1 so 2 because 3 because
4 so 5 because 6 so

VOCABULARY
1 1 nouns: flood hurricane
rain storm thunder wind
adjectives: wet
2 nouns: sun wind
adjectives: changeable dry
3 nouns: frost ice snow
adjectives: freezing

READING AND VOCABULARY
1 misty brighten up sunshine
cloudy rainy showers
heavy thunder warm
temperatures
2 b
3 1 F 2 T 3 F 4 T 5 F 6 F

SOUNDS
1 misty brighten east cloudy
Rainy possible most Wales
Northern Ireland Some heavy
warm temperatures 22°C 72°F
16°C 61°F east

LISTENING
3 Saturday: safari park - dry and
sunny
Sunday: museums - cloudy,
strong wind, rain

Lesson 37

READING
1 a flood
2 a 2 b 5 c 3 d 4 e 1

GRAMMAR
1 1 do / will be 2 fail / won't
be able 3 don't have / will
buy 4 don't get / will wait 5
drink / will ... feel 6 is /
won't go

SOUNDS
2 1 I leave 2 we listen 3
they'll learn 4 I lose 5 we
like 6 they'll look

LISTENING
1 Read the tapescript and check
your answers.

Lesson 38

SOUNDS
1 words with /k/: cassette, cat,
comb, companion, computer
words with /s/: cassette,
insurance, juice, rice, salt,
seafood, soap
words with /ʃ/: champagne,
insurance, shoes, shopping

VOCABULARY
1 1 novel 2 dog 3 juice 4
pen 5 telephone 6 soap

GRAMMAR
1 1 Which place would you go to?
2 Why would you choose that
one?
3 When would you go?
4 Who would you take with
you?
5 How would you spend your
holiday?

LISTENING
1 Speaker 1: perfect holiday
Speaker 2: perfect home
Speaker 3: perfect car
Speaker 4: perfect office
2 Read the tapescript and check
your answers.

READING AND WRITING
1 *Example answers*
1 Which books would you
read?
2 What kind of music would
you listen to?
3 What would you watch on
television?
4 Would you send any
postcards?
5 Would you take any
exercise?
6 What would you miss?

Lesson 39

GRAMMAR
1 1 went / would take 2 was or
were / would have 3 met /
would tell 4 left / would
borrow 5 would help /
wanted 6 would try / offered

SOUNDS
2 1 I'd travel 2 we do 3 I'd
take 4 I'd dance 5 you drive
6 they'd tell

Lesson 40

GRAMMAR AND READING
1 e d a c b
2 1 would 2 had 3 had
4 would 5 had 6 would
7 had 8 would
3 1 I'd cut 3 She'd 6 He'd
4 1 our friends had already left
2 her brother had become ill
3 I had tried them on
4 the guard blew his whistle
5 The party had started
6 I saw the dog
5 1 left 2 seen 3 gone 4 died
5 flown 6 lived
6 1 decided 2 had chosen
3 made 4 asked 5 had been
6 found 7 had arrived
8 hadn't taken

LISTENING
1 2
2 They had decided not to go on
motorways. Mary had an
argument with one of her
friends because he refused to
pay anything towards the fine.
She hasn't spoken to him
since.

Wordlist

The first number after each word shows the lesson in the Student's Book in which the word first appears in the vocabulary box. The numbers in *italics* show the later lessons in which the word appears again.

abroad /əbrɔːd/ 7, *11*
accept /əkˈsept/ 1
accident /ˈæksɪdənt/ 37, *40*
accountant /əˈkaʊntənt/ 12
ache /eɪk/ 29
actor /ˈæktə(r)/ 12
advertise /ˈædvətaɪz/ 7
aeroplane /ˈeərəpleɪn/ 28
airport /ˈeəpɔːt/ 8
allow /əˈlaʊ/ 7
ambulance /ˈæmbjʊləns/ 37
angry /ˈæŋgrɪ/ 32
ankle /ˈæŋkl/ 21
anniversary /ˌænɪˈvɜːsərɪ/ 32
answer /ˈɑːnsə(r)/ 1
apples /ˈæplz/ 15
architecture /ˈɑːkɪtektʃə(r)/ 10
arithmetic /əˈrɪθmətɪk/ 12
arm /ɑːm/ 21, *29*
arrest /əˈrest/ 40
arrive /əˈraɪv/ 1, *20*
art gallery /ˈɑːt gælərɪ/ 10
ask /ɑːsk/ 1
asleep /əˈsliːp/ 27
aspirin /ˈæsprɪn/ 13
atmosphere /ˈætməsfɪə(r)/ 32
attractive /əˈtræktɪv/ 17
aunt /ɑːnt/ 9
autumn /ˈɔːtəm/ 36
awful /ˈɔːfl/ 4
axe /æks/ 27

back /bæk/ 21
backpack /ˈbækpæk/ 13
bad /bæd/ 10
bag /bæg/ 6, *25*
bald /bɔːld/ 17
ballet /ˈbæleɪ/ 16
bananas /bəˈnɑːnəz/ 15
bank /bæŋk/ 5, *14, 24*
banker /ˈbæŋkə(r)/ 12
bar /bɑː(r)/ 28
barbecue /ˈbɑːbɪkjuː/ 23
basin /ˈbeɪsn/ 3
bath /bɑːθ/ 3
bathroom /ˈbɑːθruːm/ 3
be at /bi: æt/ 31
beach /biːtʃ/ 4, *10, 28, 33*
beard /bɪəd/ 17
beautiful /ˈbjuːtɪfl/ 10, *14, 17*

bed and breakfast
    /bed ən ˈbrekfəst/ 8
bed /bed/ 3
bedroom /ˈbedruːm/ 3
beef /biːf/ 15
beer /bɪə(r)/ 15
belief /brˈliːf/ 34
bicycle /ˈbaɪsɪkl/ 26
big /bɪg/ 6
bill (US) /bɪl/ 24
biology /baɪˈɒlədʒɪ/ 12
biscuit /ˈbɪskɪt/ 15, *31*
bite /baɪt/ 29
black /blæk/ 17, *19*
blonde /blɒnd/ 17
blouse /blaʊz/ 19
blue /bluː/ 19
boarding pass /ˈbɔːdɪŋ pɑːs/ 8
boat /bəʊt/ 7
body /ˈbɒdɪ/ 21
bomb /bɒm/ 37
book (v) /bʊk/ 8
border /ˈbɔːdə(r)/ 20
boring /ˈbɔːrɪŋ/ 10
borrow /ˈbɒrəʊ/ 30
bottle /ˈbɒtl/ 15
bowl /bəʊl/ 35
boy /bɔɪ/ 9
boyfriend /ˈbɔɪfrend/ 9
bread /bred/ 15
break down (v) /breɪk ˈdaʊn/ 40
break /breɪk/ 37
breakfast /ˈbrekfəst/ 2
bridge /brɪdʒ/ 32
bring /brɪŋ/ 7
broken /ˈbrəʊkən/ 6
brother /ˈbrʌðə(r)/ 9
brown /braʊn/ 17, *19*
Buddhist /ˈbʊdɪst/ 34
bunch /bʌntʃ/ 6
burglar /ˈbɜːglə(r)/ 37
burn /bɜːn/ 37
bus station /ˈbʌs steɪʃn/ 14
bus /bʌs/ 5, *28*
business class /ˈbɪznɪs klɑːs/ 8
busy /ˈbɪzɪ/ 10
butter /ˈbʌtə(r)/ 15
button /ˈbʌtn/ 37
buy /baɪ/ 5, *7*

cabbage /ˈkæbɪdʒ/ 15
cabin /ˈkæbɪn/ 8
cafe /ˈkæfeɪ/ 10, *39*
call out /kɔːl ˈaʊt/ 31
calm /kɑːm/ 18
camera /ˈkæmrə/ 13, *25*
canal /kəˈnæl/ 28
candle /ˈkændl/ 34
capital /ˈkæpɪtl/ 40
car park /ˈkɑː pɑːk/ 24
careful /ˈkeəfəl/ 18

carpet /ˈkɑːpɪt/ 3
carriage /ˈkærɪdʒ/ 26
carrots /ˈkærəts/ 15
carry /ˈkærɪ/ 7
cash /kæʃ/ 7
cassette recorder /kəˈset
    rɪˈkɔːdə(r)/ 38
casual /ˈkæʒʊəl/ 19
cat /kæt/ 38
catch fire /kætʃ ˈfaɪə(r)/ 37
cathedral /kəˈθiːdrəl/ 10
Catholic /ˈkæθəlɪk/ 34
celebrate /ˈseləbreɪt/ 23
celebration /ˌseləˈbreɪʃn/ 34
cemetery /ˈsemətrɪ/ 34
central heating
    /sentrəl ˈhiːtɪŋ/ 38
centre /ˈsentə(r)/ 33
century /ˈsentʃərɪ/ 32
chair /tʃeə(r)/ 3
champagne /ʃæmˈpeɪn/ 38
change /tʃeɪndʒ/ 5, *7, 11*
changeable /tʃeɪndʒəbl/ *36*
charm /tʃɑːm/ 14
cheap /tʃiːp/ 4, *10*
check /tʃek/ 7
check in /tʃek ɪn/ 8
check out /tʃek aʊt/ 8
cheers /tʃɪəz/ 35
cheese /tʃiːz/ 15
chemist /ˈkemɪst/ 5, *24*
chemistry /ˈkemɪstrɪ/ 12
chicken /ˈtʃɪkɪn/ 15
child /tʃaɪld/ 9
chin /tʃɪn/ 35
chips (GB) /tʃɪps/ 24, *35*
chips (US) /tʃɪps/ 24
chocolate /ˈtʃɒklət/ 31
choose /tʃuːz/ 7
church /tʃɜːtʃ/ 28
cinema /ˈsɪnəmə/ 10, *16*
city /ˈsɪtɪ/ 33
clever /ˈklevə(r)/ 18
cliff /klɪf/ 33
climate /ˈklaɪmɪt/ 10
climb /klaɪm/ 27
close /kləʊz/ 5
closing time /ˈkləʊzɪŋ taɪm/ 16
cloth /klɒθ/ 25
cloud /klaʊd/ 6
club /klʌb/ 16
coast /kəʊst/ 33
coat /kəʊt/ 6, *19*
coffee /ˈkɒfɪ/ 15
cold /kəʊld/ (adj) 4, *10, 18*
cold /kəʊld/ (n) 29
college /ˈkɒlɪdʒ/ 14
comb /kəʊm/ 38
come /kʌm/ 2
come from /kʌm frəm/ 1

come in /kʌm ɪn/ 31
companion /kəmˈpænɪən/ 38
computer /kəmˈpjuːtə(r)/ 12, *38*
concert /ˈkɒnsət/ 16
confident /ˈkɒnfɪdənt/ 18
confusing /kənˈfjuːzɪŋ/ 4
connection /kəˈnekʃn/ 8
consulate /ˈkɒnsjʊlət/ 37
conversation /ˌkɒnvəˈseɪʃn/ 14
cooker /ˈkʊkə(r)/ 3
cost /kɒst/ 20, *40*
cottage /ˈkɒtɪdʒ/ 32
cotton /ˈkɒtn/ 25
country /ˈkʌntrɪ/ 7
countryside /ˈkʌntrɪsaɪd/ 4, *33*
cousin /ˈkʌzn/ 9
crash /kræʃ/ 40
create /kriːˈeɪt/ 7
credit card /ˈkredɪt kɑːd/ 38
crisps /krɪsps/ 24
cross /krɒs/ 5
crowd /kraʊd/ 32
crowded /ˈkraʊdɪd/ 4, *10*
cup /kʌp/ 15, *31, 35*
cupboard /ˈkʌbəd/ 3
curly /ˈkɜːlɪ/ 17
currency /ˈkʌrənsɪ/ 13
curtains /ˈkɜːtnz/ 3
curved /kɜːvd/ 25
customs /ˈkʌstəmz/ 26

dancer /ˈdɑːnsə(r)/ 12
dancing /ˈdɑːnsɪŋ/ 23
dangerous /ˈdeɪndʒərəs/ 10, *37*
dark /dɑːk/ 17
date /deɪt/ 30
daughter /ˈdɔːtə(r)/ 9
dead /ded/ 34
delay /dɪˈleɪ/ 8
dentist /ˈdentɪst/ 39
departure /dɪˈpɑːtʃə(r)/ 8
desert (n) /ˈdezət/ 20, *33*
diary /ˈdaɪərɪ/ 30
die /daɪ/ 7
dining room /ˈdaɪnɪŋ ruːm/ 3
dinner /ˈdɪnə(r)/ 2
dirty /ˈdɜːtɪ/ 4, *10*
disappear /ˌdɪsəˈpɪə(r)/ 32
disco /ˈdɪskəʊ/ 16
discover /dɪsˈkʌvə(r)/ 32
disease /dɪˈziːz/ 29
dish /dɪʃ/ 35
dishwasher /ˈdɪʃwɒʃə(r)/ 3
distance /ˈdɪstəns/ 20
doctor /ˈdɒktə(r)/ 12
dog /dɒg/ 38
door /dɔː(r)/ 3, *27, 31*
double room /ˈdʌbl ruːm/ 8
draw /drɔː/ 5
dress /dres/ 19
drink /drɪŋk/ 1

drive /draɪv/ 20, *26*
driver /draɪvə(r)/ 20
driving /draɪvɪŋ/ 4
driving licence
        /draɪvɪŋ laɪsns/ 37
drown /draʊn/ 37
druggist (US) /drʌgɪst/ 24
dry /draɪ/ 36
duty-free /djuːtɪ 'friː/ 26

ear /ɪə(r)/ 21, *29*
earn /ɜːn/ 1, *11*
east /iːst/ 33
economics /iːkə'nɒmɪks/ 12
education /edʒu'keɪʃn/ 12, *22*
eggs /egz/ 15
elbow /elbəʊ/ 21, *35*
election /ɪ'lekʃn/ 22
electricity /ɪˌlek'trɪsətɪ/ 37
elegant /elɪgənt/ 14
employment /ɪm'plɔɪmənt/ 22
engineer /ˌendʒɪ'nɪə(r)/ 12
entertainment
        /ˌentə'teɪnmənt/ 10
escalator /eskəleɪtə(r)/ 28
evil /iːvl/ 32
excellent /eksələnt/ 10
exchange /ɪks'tʃeɪndʒ/ 7, *39*
excuse me /ɪk'skjuːz miː/ 39
exhaust pipe /ɪg'zɔːst paɪp/ 40
exhibition /ˌeksɪ'bɪʃn/ 16
expect /ɪk'spekt/ 39
expensive /ɪk'spensɪv/ 10
explode /ɪk'spleʊd/ 37
eyes /aɪz/ 21

face /feɪs/ 17, *21*
factory /fæktərɪ/ 10, *33*
fair /feə(r)/ 17, *23*
fall off /fɔːl ɒf/ 40
famous /feɪməs/ 14
fare /feə(r)/ 8
farm /faːm/ 27
farmland /faːmlænd/ 33
fast /faːst/ 26
fat /fæt/ 17
father /faːðə(r)/ 9
faucet (US) /fɔːsɪt/ 24
favour /feɪvə(r)/ 39
favourite /feɪvrɪt/ 32
ferry /ferɪ/ 8
field /fiːld/ 33
film /fɪlm/ 16
finally /faɪnəlɪ/ 31
finance /faɪnæns/ 22
finger /fɪŋgə(r)/ 21
finish /fɪnɪʃ/ 2
fire /faɪə(r)/ 37
fireworks /faɪəwɜːks/ 23
first class /fɜːst klaːs/ 26
fish /fɪʃ/ 15
fit /fɪt/ 18
flag /flæg/ 23
flat /flæt/ (adj) 6, *33*
flat /flæt/ (n) 3

flood /flʌd/ 36, *37*
flower /flaʊə(r)/ 6
flu /fluː/ 29
fog /fɒg/ 36
food /fuːd/ 4, *5, 10, 13*
foot /fʊt/ 21
foreign /fɒrən/ 7, *11*
forest /fɒrɪst/ 33
fork /fɔːk/ 35
formal /fɔːml/ 19, *30*
fortunately /fɔːtʃənətlɪ/ 31
freedom /friːdəm/ 22
freezing /friːzɪŋ/ 36
french fries (US)
        /frentʃ 'fraɪz/ 24
fridge /frɪdʒ/ 3
friend /frend/ 9
friendly /frendlɪ/ 4
front /frʌnt/ 31
frost /frɒst/ 36
fruit /fruːt/ 15
funny /fʌnɪ/ 18

gallery /gælərɪ/ 16
gallon /gælən/ 20
garden /gaːdn/ 3, *31*
gas /gæs/ 37
gas (US) /gæs/ 24
gas station (US)
        /gæs steɪʃn/ 20
generous /dʒenərəs/ 4
gentleman /dʒentlmən/ 39
geography /dʒɪ'ɒgrəfɪ/ 12
get /get/ 2, *5, 20*
get dressed /get drest/ 2
get on with someone /get 'ɒn
        wɪð sʌmwʌn/ 30
get out /get aʊt/ 31
get up /get ʌp/ 2
girl /gɜːl/ 9
girlfriend /gɜːlfrend/ 9
give up /gɪv ʌp/ 7
glass /glaːs/ 15, *25*
glasses /glaːsɪz/ 17
glue /gluː/ 25
go /gəʊ/ 1
go into /gəʊ ɪntuː/ 31
go shopping /gəʊ ʃɒpɪŋ/ 2
go to sleep /gəʊ tə sliːp/ 2
good /gʊd/ 10
give /gɪv/ 1
good-looking /gʊd lʊkɪŋ/ 17
government /gʌvnmənt/ 22
grandfather /grændfaːðə(r)/ 9
grandmother /grændmʌðə(r)/ 9
grapes /greɪps/ 15
grass /graːs/ 4, *27, 32*
great /greɪt/ 4
green /griːn/ 19
grey /greɪ/ 19
ground floor /graʊnd flɔː(r)/ 37
grounds /graʊndz/ 32
guard /gaːd/ 26, *32, 40*
guide book /gaɪd bʊk/ 13
gun /gʌn/ 4, *37*

hair /heə(r)/ 17, *21*
hairdryer /heə(r)draɪə(r)/ 25
ham /hæm/ 15
hammer /hæmə(r)/ 27
hand /hænd/ 35
handbag /hændbæg/ 13
hangover /hæŋəʊvə(r)/ 29
happen /hæpən/ 39
harbour /haːbə(r)/ 8
hard /haːd/ 25
hat /hæt/ 19
have /hæv/ 2
have a shower/bath
        /hæv ə 'ʃaʊə(r)/baːθ/ 2
head /hed/ 17, *21, 29*
heart /haːt/ 29
heavy /hevɪ/ 6, *25*
high /haɪ/ 25
high blood pressure
        /haɪ 'blʌd preʃə(r)/ 29
high street /haɪ striːt/ 24
highlight /haɪlaɪt/ 34
highway /haɪweɪ/ 20
hill /hɪl/ 20, *27, 33*
hilly /hɪlɪ/ 33
history /hɪstrɪ/ 12
hold /həʊld/ 5
hold on /həʊld ɒn/ 30
holiday /hɒlədeɪ/ 7, *23*
home /həʊm/ 2, *3*
hospital /hɒspɪtl/ 14, *28*
hot /hɒt/ 4
hotel /həʊ'tel/ 7, *28*
house /haʊs/ 3, *11*
housing /haʊzɪŋ/ 22
humid /hjuːmɪd/ 36
hurricane /hʌrɪkən/ 36
hurry /hʌrɪ/ 27
husband /hʌzbənd/ 9

ice /aɪs/ 36
ice cream /aɪskriːm/ 35
ill /ɪl/ 29
imaginative /ɪ'mædʒɪnətɪv/ 18
industrial /ɪn'dʌstrɪəl/ 33
industry /ɪndəstrɪ/ 33
inflation /ɪn'fleɪʃn/ 22
informal /ɪn'fɔːml/ 30
injured /ɪndʒəd/ 37
insect /ɪnsekt/ 4, *29*
insurance /ɪn'ʃʊərəns/ 38
intelligent /ɪn'telɪdʒənt/ 18
interesting /ɪntrəstɪŋ/ 10, *18*
interval /ɪntəvl/ 16
introduce /ˌɪntrə'djuːs/ 7
invade /ɪn'veɪd/ 32
island /aɪlənd/ 33

jacket /dʒækɪt/ 19
jam /dʒæm/ 35
jeans /dʒiːnz/ 19
jet lag /dʒet læg/ 29
jewellery /dʒuːəlrɪ/ 38
jobs /dʒɒbz/ 11
jokes /dʒəʊks/ 14
journalist /dʒɜːnəlɪst/ 12

journey /dʒɜːnɪ/ 40
juice /dʒuːs/ 15
jungle /dʒʌŋgl/ 33

kill /kɪl/ 37
kilo /kiːləʊ/ 15, *37*
kind /kaɪnd/ 17, *18*
king /kɪŋ/ 23
kitchen /kɪtʃɪn/ 3
knee /niː/ 21
knife /naɪf/ 35
knock /nɒk/ 27
know /nəʊ/ 1

ladder /lædə(r)/ 27
lake /leɪk/ 33
lamb /læm/ 15
lamp /læmp/ 3
land /lænd/ 8
lane /leɪn/ 26
language /læŋgwɪdʒ/ 11, *12*
lap /læp/ 35
large /laːdʒ/ 10
law and order
        /lɔːr ənd 'ɔːdə(r)/ 22
law courts /lɔː kɔːts/ 14
lawn /lɔːn/ 32
lazy /leɪzɪ/ 18
lead (v) /liːd/ 7
learn /lɜːn/ 11
leather /leðə(r)/ 25
leave /liːv/ 2, *7, 20*
left-wing /left 'wɪŋ/ 14
leg /leg/ 21, *29*
lend /lend/ 30
lettuce /letɪs/ 15
level crossing /levl 'krɒsɪŋ/ 28
library /laɪbrərɪ/ 14, *28*
lift /lɪft/ 8, *28*
light /laɪt/ 25
lightning /laɪtnɪŋ/ 36
liquid /lɪkwɪd/ 25
literary /lɪtərərɪ/ 14
live /lɪv/ 1
lively /laɪvlɪ/ 14
living room /lɪvɪŋ ruːm/ 3
loaf /ləʊf/ 15
local /ləʊkl/ 14
lock /lɒk/ 27
long /lɒŋ/ 17, *25*
look at /lʊk æt/ 5
look for /lʊk fɔː/ 31
lose one's way
        /luːz wʌnz weɪ/ 32
low /ləʊ/ 6, *25*
lower (v) /ləʊə(r)/ 27
luggage /lʌgɪdʒ/ 8
lunch /lʌntʃ/ 2
lung /lʌŋ/ 29
luxury /lʌkʃərɪ/ 7

machine /mə'ʃiːn/ 25
main street (US) /meɪn striːt/ 24
man /mæn/ 9, *17*
manners /mænə(r)z/ 30
map /mæp/ 13

Wordbank

Use the categories below to help you organise new vocabulary. Try and write each new word in at least two different categories. You may also like to write down words which often go with the new vocabulary items.

character	clothes	countries and nationalities
crime and justice	customs and traditions	daily life
days, months, seasons	education	environmental issues
family and friends	food and drink	geographical features and locations
health and physical feelings	house and home	language learning
leisure interests	the media	parts of the body
personal information	personal possessions	physical appearance
politics, government and society	religion	shops and shopping
social situations	town features and facilities	transport
travel	work	weather